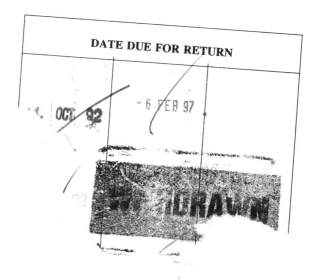

THE ECONOMICS OF ENVIRONMENT

Papers from Four Nations

THE ECONOMICS OF ENVIRONMENT

Papers from Four Nations

Edited by

PETER BOHM and ALLEN V. KNEESE

Originally published in

The Swedish Journal of Economics

MACMILLAN
ST. MARTIN'S PRESS

The essays are reprinted from *The Swedish
Journal of Economics,* vol. 73, no. 1 (March 1971)

Published 1971 by
THE MACMILLAN PRESS LTD
London and Basingstoke
*Associated companies in New York Toronto
Dublin Melbourne Johannesburg and Madras*

Library of Congress catalog card no. 73-178244

SBN 333 13276 9

Printed in Great Britain by
COMPTON PRINTING LTD
London and Aylesbury

Contents

Part I
PRELIMINARIES

Introduction

by Peter Bohm (Stockholm School of Economics)
and Allen V. Kneese (Resources for the Future, Inc.).

Generations of economists recognized the existence of external or "social" costs like those resulting from pollution, associated with private production and consumption activity. They also saw that these phenomena were connected with "common property"—that is instances in which, for institutional or technical reasons, valuable attributes of economic assets were not effectively reduced to private ownership. Therefore these assets could not enter into the exchange process basic to the efficiency of market systems. The attention of a small but penetrating group of theorists laid a firm conceptual base for developing an economic theory reaching beyond analysis of the positive and normative aspects of market exchange. But it is fair to say that until very recently most economists, explicitly or implicitly, regarded the common property-externality-congestion complex of phenomena as minor ripples in an otherwise smoothly flowing stream of exchange. There were, of course, exceptions. Pigou in his *Economics of Welfare*, Kapp in his *Social Costs of Private Enterprise*, and Baumol in his *Welfare Economics and the Theory of the State* sent early and clear signals that important aspects of reality were not being given their due.

Nevertheless it is not until the last five years or so that the economic literature has begun to recognize explicitly that externality phenomena are a pervasive element in highly developed exchange economies and that a body of research has developed aimed at supplying solid help in managing the common property assets of society.

This book is at least a small landmark in this development. It brings together papers by a group of economists, most of whom are internationally known, which reflect the serious attention environmental economics is getting. Represented here are the first publications on environmental economics by several of these prominent scholars.

This book should first of all be regarded as an example of the response from economists to the challenge of environmental degradation. This challenge to economic science is twofold. There is the immediate need for simple presentations of important economic facts to be conveyed to politicians and other readers who are not so well trained in environmental economics or economics in general. There is also the need for new theoretical approaches and developments and for empirical studies in this field. Both categories are represented in this volume.

The book is divided into four parts. The first consists of this introduction and a paper by Kneese. The latter is strictly for the purpose of providing informational background to those economists and other readers who may not be highly informed about the substantive aspects of environmental pollution. The papers in the other parts address more distinctly economic questions in the areas of theory, methodology, and applications.

Part II consists of papers by d'Arge, Dahmén, Baumol and Oates, and Mishan. These papers are mostly concerned with the theory of economic policy in regard to environmental degradation. In them one sees most clearly how the fact of pervasive externalities is being incorporated into the theory of general equilibrium and into growth theory. These authors have done an excellent job of presenting deep and difficult matters most straightforwardly.

The papers in Part III are by Malinvaud, Bohm, Mäler, Lave and Seskin. They all deal with the problem of measuring the value of collective goods and bads—a problem which is central to rational economic policy in this area. They carry us far beyond the assertion, which has become almost a cliche in economic theory, that people cannot be induced to reveal their preferences for collective goods.

The fourth and final part of the book deals with operational models for environmental management and consists of a paper by Russell. He has built an operational model of an industrial plant which takes into account both the internal and external costs and returns of the operation. He shows how this construct fits into a large-scale model of residual materials management in a region. This pioneering methodology is now being tested in an empirical application in the Delaware Estuary region of the U.S.A.

We do not go further in describing the contents of the book since they are best described by the authors themselves. Thus, the reader is referred to the introductions or summaries presented in the beginning of their respective essays. As should be apparent already from these introductory presentations, economic science has by now tried—perhaps contrary to what the public opinion believes —to tackle the problems of environmental degradation on a wide front and with a variety of scientific methods.

The reader of this book will get a view of a profession rethinking, extending, and revising its concepts, and finding new applications for them. These essays and other current discussions in the profession, for example those about national accounting—incorporate collective goods and bads, are reminiscent of the constructive ferment in the profession when the Keynesian revolution was in progress. We are witnessing a piece of intellectual history in the making.

BACKGROUND FOR THE ECONOMIC ANALYSIS OF ENVIRONMENTAL POLLUTION

*Allen V. Kneese**

Resources for the Future, Inc., Washington, D.C., USA

Summary

Relevant and useful work on economic aspects of environmental problems requires some knowledge of the scientific and technological aspects of these problems. This article provides a summary of the latter. Global effects of environmental pollution, like effects on climate and large ecological systems, are reviewed first. Then, smaller-scale but severe "regional" problems are addressed. These include air and water pollution and solid wastes disposal. Sources and type of pollutants are identified, their effects on nature and human receptors are discussed, and available control technologies are reviewed. Finally, a more coherent view of the whole set of problems is provided by means of a "materials balance" approach.

Introduction

Economic theory has provided a conceptual structure indispensable for understanding contemporary environmental problems and for formulating effective and efficient policy approaches toward them. Concepts like external diseconomies and public goods provide enormously useful insights. But economic theorizing and research that take place without being well informed about the substantive character of the problems under study is in danger of being somewhat arid because of extreme abstraction or of expending scarce energy and talent in the pursuit of relatively unimportant matters. The objective of the present essay is to provide an introduction to the substantive aspects of one of the major environmental problems facing both developed and some developing economies—environmental pollution.

Environmental pollution has existed for many yers in one form or another. It is an old phenomenon,[1] and yet in its contemporary forms it seems to have

---

* I wish to thank my colleagues Blair Bower, Clifford Russell, and Walter Spofford for assistance in the preparation of this paper.
[1] Many accounts attest that severe environmental degradation has existed for a long time in the western countries. In fact, the immediate surroundings of most of mankind in this part of the world were much worse a century ago than they are now. The following account of statements from an address of Charles Dickens may be interesting in this connection, especially to those who know contemporary London: "He knew of many places in it [London] unsurpassed in the accumulated horrors of their long neglect by the

crept up on governments and even on pertinent professional disciplines such as biology, chemistry, most of engineering,[1] and, of course, economics. A few economists, such as Pigou, wrote intelligently and usefully on the matter a long time ago, but generally even that subset of economists especially interested in externalities seems to have regarded them as rather freakish anomalies in an otherwise smoothly functioning exchange system. Even the examples commonly used in the literature have a whimsical air about them. We have heard much of bees and apple orchards and a current favorite example is sparks from a steam locomotive—this being some eighty years after the introduction of the spark arrester and twenty years after the abandonment of the steam locomotive.

Moreover, air and water continued until very recently to serve the economist as examples of free goods. A whole new set of scarce environmental resources presenting unusually difficult allocation problems seems to have appeared on the scene with the profession having hardly noticed. Fortunately, this situation is changing fast and much good work is appearing in the current economics literature as the papers in this issue amply demonstrate.

Substantial and thoughtful attention from economists is especially needed because the economic and institutional sources of the problem are either neglected or thoroughly misunderstood by most of those currently engaged in the rather frantic discussion of it.

I. "Global" Problems

I will begin this discussion of the substantive aspects of pollution problems by concentrating first on those problems, or potential problems, which affect the entire planet. Thereafter I will focus on "regional" problems. By regional I mean all those other than global. One must use a word like regional rather than terms pertaining to political jurisdictions such as nations, states, or cities because the scale of pollution resulting from the emissions of materials and energy follows the patterns, pulses, and rhythms of meteorological and hydrological systems rather than the boundaries of political systems—and therein lies one of the main problems.

The global problems to be discussed here pertain largely to the atmosphere because the marks of man have already been seen on that entire thin film of

---

dirtiest old spots in the dirtiest old towns, under the worst old governments of Europe." He also said that the surroundings and conditions of life were such that "infancy was made stunted, ugly and full of pain—maturity made old—and old age imbecile." These statements are from *The Public Health a Public Question: First Report of the Metropolitan Sanitary Association*, address of Charles Dickens, Esq., London,1850. Great achievements in the elementary sanitation of the close-in environment have been made as well as impressive gains in public health. The distinguishing feature of contemporary environmental pollution seems to be the large-scale and subtle degradation of common property resources. This point is developed in the text below.

[1] There has been a relatively small group of sanitary engineers that has given close attention to environmental problems for a long time. I am here referring to the mainstream of work in these professions.

life-sustaining substance. It seems to have come as something of a shock to the natural science community that man not only can, *but has*, changed the chemical composition of the whole atmosphere.[1] Other large-scale problems, or potential problems, particularly those related to the "biosphere" will be discussed more briefly.

Before proceeding to what may be real global problems, it will be desirable to dispose of one red herring. One of the spectres raised by the more alarmist school of ecologists is that man will deplete the world's oxygen supply by converting it into carbon dioxide in the process of burning fossil fuels for energy. This idea has now been thoroughly discredited by two separate pieces of evidence. The first is measurement of changes in the oxygen supply over a period of years. There is currently *one* monitoring station in the world whose objective it is to identify long-term changes in the atmosphere. The station is operated at a high elevation in Hawaii by the U.S. Weather Bureau. Observations there have shown the oxygen content of the atmosphere to be remarkably stable. The other piece of evidence—perhaps more persuasive—is in the form of a "gedankenexperiment". If one burns, on paper, the entire known world supply of fossil fuels and all the present plant biomass, the impact on the oxygen supply is to reduce it by about 3 %. This is much too small to be noticed in most areas of the earth.

Potentially real effects on the atmosphere and climate are thought to be connected with changes in carbon dioxide and particulate matter (including aerosols) in the atmosphere, petroleum in the oceans, waste energy rejection to the atmosphere, and the widespread presence of toxic agents in the coastal waters and oceans. I will discuss each of these briefly in turn.

The production of carbon dioxide is an inevitable result of the combustion of fossil fuels. In contrast to O_2, the relative quantity of CO_2 in the atmosphere has increased measurably. The possible significance of this is that CO_2 absorbs infrared radiations and therefore an increasing concentration of it in the atmosphere would tend to cause the surface of the earth to rise in temperature.[2] Some estimates have put the possible increase in CO_2,[3] if present rates of increase in the combustion of fossil fuel continue, at about 50 % by the end of the century. An increase of this amount could raise the world's mean surface temperature several degrees with attendant melting of ice caps, inundation of seacoast cities, and undesirable temperature increases in densely inhabited areas. Estimates made in the summer of 1970 suggest, however, that the CO_2 increase will only be about 20 % by the end of the century with lesser

[1] The fullest discussion of the range of problems considered in this section will be found in *Man's Impact on the Global Environment: Assessment and Recommendations for Action.* M.I.T. Press, Cambridge (Mass.), 1970.
[2] Most incoming radiation is in the form of visible light, while most outgoing energy is in the form of infrared radiation.
[3] Conservation Foundation, *Implications of Rising Carbon Dioxide Content of the Atmosphere*, New York, 1963.

potential effects on climate.[1] The difference in estimates is accounted for by the newly recognized fact that less of the CO_2 generated by combustion is staying in the atmosphere than was previously supposed. Apparently one or more of the "sinks" for CO_2 is responding to the increased concentration, or possibly even a third force is leading to greater absorption. The main sinks for CO_2, or more specifically carbon, are solution in the oceans and conversion by the flora of the earth. Perhaps carbon is limiting to growth in some of these plant populations, and they are responding to its increased availability from the atmosphere.[2] Another possibility is less reassuring. Somewhat anomalously the mean temperature of the earth's surface has been falling over the past couple of decades according to Weather Bureau observations. As is true of many gases, the solubility of CO_2 in water increases when water temperature falls. Maybe that's where some of the CO_2 went.

But this brings us to another possible effect of man's activity on world climate. Some meteorologists—especially Bryson[3]—believe that man's industrial and agricultural activities are causing the world to cool off. The suspected mechanism is an increase in particulates and aerosols which, they think, are increasing the earth's albedo (ability to reflect incoming solar radiation). Farming and other activities in arid areas and the combustion of fuels send immense amounts of particulates and fine water vapors (aerosols) into the atmosphere each day. This is an undisputed fact although observations indicate that a worldwide increase in particulate matter cannot yet be identified.[4] What is in dispute is the effect of the man-generated increase. Some not only believe this effect is significant, but that it may be sending us rather rapidly toward an ice age—perhaps the final ice age resulting in a perpetually frozen planet. Other factors might lead in the same direction as we shall see subsequently. The freeze-up hypothesis is, however, disputed by those other meteorologists who regard it as important to recognize the difference between particulates of different types and at different elevations in the atmosphere. Mitchell has pointed out that there has been a large amount of volcanic activity in recent years which has deposited great quantities of particulates at high elevations in the atmosphere.[5] The net effect of these is fairly clearly to reflect more energy away from the earth, and this could well be responsible for the observed temperature decline. On the other hand, he points out that particulates deposited at relatively low altitudes, such as those generated by man,

---

[1] *Man's Impact on the Global Environment* ... etc., op. cit.

[2] With higher concentrations of CO_2 in the atmosphere, one would naturally expect some increase in absorption by the oceans, since CO_2 solubility is a function of the partial pressure of CO_2.

[3] R. A. Bryson and J. T. Peterson: Atmospheric aerosols: Increased concentrations during the last decade. *Science 162*, 3849 (Oct. 1968), pp. 120–21.

[4] It should be noted that this is a somewhat controversial conclusion.

[5] J. M. Mitchell, Jr.: A preliminary evaluation of atmospheric pollution as a cause of the global temperature fluctuation of the past century. In *Global Effects of Environmental Pollution* (ed. Singer) pp. 97–112. Reidel, Holland, 1970.

could well have the reverse effect because they reflect energy back toward earth as well as away from it. Mitchell's calculations tend to show that the former effect outweighs the latter.[1] Thus, when the effect of the volcanic particulates wears off over a period of years, the lower altitude particulates could begin to reinforce CO_2 as a factor leading to rising world temperatures.

An additional, and possibly reinforcing factor, is the release of energy to the atmosphere due to the energy conversion activities of man. A large proportion of the energy from fuels man uses is transferred directly to the atmosphere —as, for example, the energy converted in automobile engines. Another large proportion is initially tranferred to water—as, for example, when condensers in electric power plants are cooled with water. But this too is rather quickly rejected to the atmosphere by induced evaporation in watercourses or wet-cooling towers. Essentially, all of the energy converted from fuels is transferred to the atmosphere as heat. Because this is so, it is possible to make a rather precise estimate of this transfer by calculating the energy value of the fuels used in the world. On this basis there is an average emission of about 5.7×10^{12} watts of energy from human conversion.[2] What does one make of such a monstrous number? More understandable perhaps is the statement that this is about 1/15 000 of the absorbed solar flux. That doesn't seem like much, but another important element in the picture is the fact that energy conversion is a rapidly growing activity all over the world. The most spectacular example is conversion to electric power which in the United States has been proceeding at a doubling time of ten years and even faster in one or two other large economies. Worldwide energy conversion (by far the largest proportion of which is from fossil fuels) as a whole has been proceeding at a growth rate of about 4 % a year. If we project this growth rate for 130 years, we will reach a rate of energy rejection of about 1 % of the absorbed solar flux. This is enough, some meteorologists believe, to have a substantial effect on world climate. If we proceed at the 4 % growth rate for another 120 years, we will have reached 100 % of the absorbed solar flux. This would be a total disaster. The resulting mean increase in world temperature would be about 50°C—a condition totally unsuitable for human habitation. We will never reach such a situation, but the important question is what circumstances will prevent us from so doing.

If one is given to apocalyptic visions, he can readily imagine a situation in which CO_2, particulates, and energy conversion reinforce each other and will, after a short reprieve from the volcanoes, make the earth into a kind of mini-hell.

But other things may happen too. For one thing, we are annually spilling on the order of 1.5×10^6 tons of oil directly into the oceans with perhaps another 4×10^6 tons being delivered by terrestrial streams. This may be enough "oil on

---

[1] Personal communication to William Frisken.

[2] W. R. Frisken, "Extended Industrial Revolution and Climate Change," unpublished report prepared while he was a visiting scholar at RFF, July, 1970.

troubled waters", some scientists believe, to smooth the sea surface sufficiently to cause its reflectivity to be increased significantly.[1] Again, the associated albedo effect would tend to cause cooling. But at the same time the reduction in the atmosphere–ocean interface would tend to diminish the absorption of CO_2 and thus possibly tend toward a warming condition.

And then there is the matter of the SST. The European Concorde and the Russian SST have already flown, and it looks as though the U.S. version will also fly one day. Aside from the major question of sonic booms, the emissions from SSTs may have substantial effects on the upper atmosphere. SSTs would fly at 65 000–70 000 feet and the atmosphere is very different up there. It is extremely dry and the layers at that elevation do not seem to mix much with the lower atmosphere. Five hundred SSTs might be in operation by the mid-1980s. If these were the American type, their emissions might cause an increase of water vapor in the upper atmosphere of 10 % globally and possibly 60 % over the North Atlantic where most of the flights would occur.[2] This could give rise to large-scale formations of very persistent cirrus cloud. Furthermore, the emissions of soot, hydrocarbons, nitrogen oxides, and sulfate particles could cause stratospheric smog. The effects of all this would be somewhat uncertain but presumably not unlike those produced by particulates deposited into the upper atmosphere by volcanoes—in other words, increased albedo and consequent cooling at the earth's surface.

A final category of substances of possibly global significance are persistent organic toxins. DDT is a good example of these and has been found in living creatures all over the world. How it got to remote places like the Antarctic is still somewhat mysterious, but apparently substantial amounts are transmitted through the atmosphere as well as through the oceans. Aside from possible large-scale effects on ecological systems, these persistent toxins could affect the O_2–CO_2 balance by poisoning the phytoplankton which are involved in one of the important CO_2–O_2 conversion processes. We do not know whether this is happening or not.

Clearly, we are operating in a context of great uncertainty. Equally clearly, man's activities now and in the relatively near-term future may affect the world's climatic and biological regimes in a substantial way. It seems beyond question that a serious effort to understand man's effects on the planet and to monitor those effects is indicated. Should we need to control such things as the production of energy and CO_2 in the world, we will face an economic and political resource allocation problem of unprecedented difficulty and complexity.

The discussion of global effects of pollution was necessarily somewhat

---

[1] Oil also may affect phytoplankton and other species directly.
[2] L. Machta, "Stratospheric Water Vapor," a working paper written for the 1970 Summer Study on Critical Environmental Problems, sponsored by the Massachusetts Institute of Technology and held at Williams College in Williamstown, Massachusetts, July 1970.

speculative, but now we turn to problems on a less grand scale. These regional problems are clear and present. A discussion is first presented under the traditional categories of waterborne, airborne, and solid residuals.[1] In the final section, I point explicitly to the interdependencies among these residuals streams and the implications of this for economic analysis. Unfortunately, most of the numbers given are from the United States. This is because I am simply not familiar with the data from other countries. The relationships in the United States may, however, be reasonably representative of those found in other industrialized countries.

II. Waterborne Residuals

Degradable residuals

A somewhat oversimplified but useful distinction for understanding what happens when residuals are discharged to watercourses is between *degradable* and *non-degradable* materials. The most widespread and best known degradable residual is domestic sewage, but, in the aggregate, industry produces greater amounts of degradable organic residuals almost all of which is generated by the food processing, meat packing, pulp and paper, petroleum refining, and chemicals industries. Some industrial plants are fantastic producers of degradable organic residuals: a single uncontrolled pulp mill, for example, can produce wastes equivalent to the sewage flow of a large city.

When an effluent bearing a substantial load of degradable organic residuals is expelled into an otherwise "clean" stream, a process known as "aerobic degradation" begins immediately. Stream biota, primarily bacteria, feed on the wastes and break them down into their inorganic forms of nitrogen, phosphorous, and carbon, which are basic plant nutrients. In the breaking down of degradable organic material, some of the oxygen which is dissolved in any "clean" water is utilized by the bacteria. But this depletion tends to be offset by reoxygenation which occurs through the air–water interface and also as a consequence of photosynthesis by the plants in the water. If the waste load is not too heavy, dissolved oxygen in the stream first will drop to a limited extent (say, to 4 or 5 parts per million from a saturation level of perhaps 8–10 ppm, depending upon temperature) and then rise again. This process can be described by a characteristically shaped curve or function known as the "oxygen sag". The differential equations which characterize this process were first introduced by Streeter and Phelps in 1925 and are often called the Streeter-Phelps equations.

If the degradable organic residual discharged to a stream becomes great enough, the process of degradation may exhaust the dissolved oxygen.

---

[1] Due to limitations of space, I will concentrate on material residuals as sources of environmental pollution. There is some discussion of energy residuals—especially where they interact in important ways with material residuals. Noise, an important energy residual, is not treated at all. A good introductory discussion of noise can be found in chapter 1 of the *Handbook of Noise Control* (ed. Cyril M. Harris). McGraw-Hill, New York, 1957.

In such cases, degradation is still carried forward but it takes place anaerobically, that is, through the action of bacteria which do not use free oxygen but organically or inorganically bound oxygen, common sources of which are nitrates and sulphates. Gaseous by-products result, among them carbon dioxide, methane, and hydrogen sulfide.

Water in which wastes are being degraded anaerobically emits foul odors, looks black and bubbly, and aesthetically is altogether offensive. Indeed, the unbelievably foul odors from the River Thames in mid-nineteenth century London caused the halls of Parliament to be hung with sheets soaked in quicklime and even induced recess upon occasion when the reek became too suffocating. So extreme a condition is rarely encountered nowadays, although it is by no means unknown. For example, a large lake near São Paulo, Brazil, is largely anaerobic, and most of the streams in the Japanese papermaking city Fuji are likewise lacking in oxygen. Other instances could be mentioned. But levels of dissolved oxygen low enough to kill fish and cause other ecological changes are a much more frequent and widespread problem.

High temperatures accelerate degradation. They also decrease the saturation level of dissolved oxygen in a body of water. So a waste load which would not induce low levels of dissolved oxygen at one temperature may do so if the temperature of the water rises. In such circumstances, heat may be considered a pollutant. Moreover, excess heat itself can be destructive to aquatic life. Huge amounts of heat are put into streams by the cooling water effluents of electric power plants and industry.

There is, in fact, increased concern about the impacts of heat residuals, particularly from power generation, in the face of the incessantly increasing demand for electric power already described and the development of nuclear power, the present "generation" of which requires more heat disposal per kwh generated than fossil fuel plants. Increasing use of cooling towers has been one response to this situation. But the use of cooling towers represents basically a transfer of the medium into which to reject the residual heat energy, that is, to the air instead of temporarily to the water. One author in the United States has discussed some aspects of what would happen over the central region of the United States under the alternative procedure, i.e., use of once-through cooling with discharge of waterborne heat to the main streams of the area, the Missouri and the Mississippi. About 540 million kilowatts of fossil fuel burning capacity are assumed installed and operating in this region by the year 2000. He writes:

Imposing the requirement of at least 10 miles separation between stations and noting that such a generating capacity will raise the water temperature by about 20 deg F, we find approximately 3 000 miles of river spreading over the central region of the United States with a temperature 20 deg F higher than normal.[1]

---

[1] S. M. Greenfield: *Science and the Natural Environment of Man*, p. 3, RAND Corporation, California, February 17, 1969.

Of course the ecological effects which would accompany such a large-scale heat discharge to our streams can only be speculated about at this time. If there were a substantial discharge of degradable organic residuals to these streams at the same time, they would almost certainly become anaerobic in the summer time. The freshwater life forms we are accustomed to would be lost.

A conventional sewage treatment plant processing degradable organic residuals uses the same biochemical processes which occur naturally in a stream, but by careful control they are greatly speeded up. Under most circumstances, standard biological sewage treatment plants are capable of reducing the BOD (biochemical oxygen demand) in waste effluent by perhaps 90 %. As with degradation occurring in a watercourse, plant nutrients are the end-product of the process.

Stretches of streams which persistently carry less than 4 or 5 ppm of oxygen will not support the higher forms of fish life. Even where they are not lethal, reduced levels of oxygen increase the sensitivity of fish to toxins. Water in which the degradable organic residuals have not been completely stabilized is more costly to treat for public or industrial supplies. Finally, the plant nutrients produced by bacterial degradation of degradable organic residuals, either in the stream or in treatment plants, may cause algae blooms. Up to a certain level, algae growth in a stream is not harmful and may even increase fish food, but larger amounts can be toxic to fish, produce odors, reduce the river's aesthetic appeal, and increase water supply treatment problems. Difficulties with algae are likely to become serious only when waste loads have become large enough to require high levels of treatment. Then residual plant nutrient products become abundant relative to streamflow and induce excessive plant growth.

Problems of this kind are particularly important in comparatively quiet waters such as lakes and tidal estuaries. In recent years certain Swiss and American lakes have changed their character radically because of the buildup of plant nutrients. The most widely known example is Lake Erie, although the normal "eutrophication" or aging process has been accelerated in many other lakes. The possibility of excessive algae growth is one of the difficult problems in planning for pollution control—especially in lakes, bays, and estuaries, for effective treatment processes today carry a high price tag.

In the United States, currently, BOD discharges by industry are apparently about twice as large as by municipalities. How fast BOD discharges grow depends on how effectively industrial wastes are controlled and municipal wastes treated. If current rates continue, BOD may grow about $3\frac{1}{2}$ % per year with plant nutrient discharges growing even faster.

Bacteria might also be included among what we have called the degradable pollutants since the enteric, infectious types tend to die off in watercourses, and treatment with chlorine or ozone is highly effective against them. Because of water supply treatment, the traditional scourges of polluted water—

typhoid, paratyphoid, dysentery, gastroenteritis—have become almost unknown in advanced countries. One might say that public concern with environmental pollution peaked early in this century with the rapid spread of these diseases. But public health engineers were so successful in devising effective water supply treatment that attention to water pollution lapsed until its recent upsurge.

A model for the analysis of bio-degradable residuals

I turn here to a brief description of a model which was used to analyse water quality improvement alternatives in the Delaware estuary area. I will linger over this model a while because it is an ingenious linking of an economic optimization (linear programming model) and a model of the residuals transport and degradation phenomena in the estuary. Analogous models have now been used for the economic analysis of water quality management problems in several other instances. What is illustrated here is one way of linking a diffusion model to an economic optimization model. Other ways have also been developed.[1]

The estuary was divided into 30 reaches, and the Streeter-Phelps oxygen balance equations previously mentioned were adapted and applied to these interconnected segments. This led to a system of linear first-order differential equations. The transfer functions, which related the change in segment i to an input of waste in segment j,[2] fortunately simplify to a set of linear relationships if steady-state conditions are assumed, i.e., if it is assumed that the discharge rate is constant and temperature and river flow are taken as parameters. In fact, the transfer functions can be represented by a matrix.[3]

This result is very fortunate because it means that these coefficients can readily be incorporated into a linear set of constraints which fit the linear programming format quite straightforwardly.

Assume that the watercourse consists of m homogeneous segments and c_i represents the improvement in water quality required to meet the D.O. target in segment i. The target vector c of m elements can be obtained by changes

---

[1] See, for example, the papers by K.-G. Mäler and C. S. Russell, below.

[2] The transfer relations are fairly complicated because the degradation of degradable organic residuals in the watercourse consumes D. O., whereas aeration, or movement of oxygen across the interface between air and water, tends to counteract this effect. Important variables that affect the oxygen balance for a given waste discharge are temperature and various characteristics of the watercourse.

[3] This result is explained by Sobel as follows: "The transfer function for D. O. response in segment i due to an input of frequency ω in segment j is:
$\Phi_{ij}(\omega) = a_{ij}(\omega) \exp z\theta_{ij}(\omega)$
where $\theta_{ij}(\omega)$ is the phase shift, $a_{ij}(\omega)$ is the amplitude attenuation, and $z = \sqrt{-1}$. It can be shown that $\theta_{ij}(\omega)$ is an arctangent function whole value $\to 0$ as $\omega \to 0$; as $\omega \to 0$, $\Phi_{ij} \to a_{ij}(0)$. $a_{ij} \equiv a_{ij}(0)$ is the D.O. response in segment i per unit of steady-state input in segment j." Sobel attributes the development of this result to V. V. Solodovnikov, *Introduction to the Statistical Dynamics of Automatic Control Systems* (Dover Publications, New York, 1960). The discussion in this section is based on Matthew J. Sobel, "Water Quality Improvement Programming Problems," *Water Resources Research*, Fourth Quarter, 1965.

of inputs to the water resource from combinations of the m segments. Define a program vector $x = (x_j, x_1, ..., x_n)$ in which the values of x refer to the volume of waste discharges in each of the estuary reaches. In a feasible solution, these values represent the waste discharges at the various points which meet the target vector c. This vector generates D.O. changes through the mechanism of the constant coefficients of the linear system already described—a_{ij} = D.O. improvement in segment i per unit of x_j, $i = 1, ..., m$; $j = 1, ..., n$; and, of course, $x_j > 0$.

If we let A be the $m \times n$ matrix of a_{ij} coefficients, then Ax is the vector of D.O. changes corresponding to x.

Now, recalling that c is the vector of target improvements, we have two restrictions on x, namely, $Ax \geqslant c$ and $x \geqslant 0$. The reader will have noticed that mathematically these are sets of linear constraints such as those found in a standard linear program. All we need is an objective function to complete the problem. Let d be a row vector where d_j = unit cost of x_j, $j = 1, ..., n$. Notice that this assumes linear (or piece-wise linear) cost functions.[1] We can now write the problem as a standard linear program,[2]

min dx

s.t. $Ax \geqslant c$

$x \geqslant 0$

Of course the transfer coefficients (a_{ij}) relate to a steady-state condition and to specified conditions of stream flow and temperature. Thus the model turns out to be totally deterministic, and the variability of conditions is handled in this analysis by assuming extreme conditions usually associated with substantial declines in water quality. This model was used to analyze the costs of alternative treatment strategies for water quality improvement in the estuary.

Similar models can be devised for the other waterborne residuals to be discussed shortly. Usually the transfer coefficient matrices are much simpler to calculate for non-degradable residuals. In fact, such models can be built for a wide variety of residuals which are degraded or diluted in nature. They go under the general title of "diffusion models" and play an important role in the contemporary economic analysis of environmental quality problems. For example, such models are used in the Russell-Spofford procedure discussed in a paper in this volume.[3]

---

[1] Programs with linear constraints and non-linear objective functions can usually be solved if the non-linear function is not too complicated. So this condition would not necessarily have to hold.

[2] The actual programs needed to solve the problem encountered in the Delaware Estuary were somewhat more complicated. The reader interested in the details should consult the paper by Sobel, op. cit.

[3] See C. S. Russell, "Models for the Investigation of Industrial Response to Residuals Management Actions," in this volume.

Non-degradable pollution

BOD serves as a good indicator of pollution where one aspect is concerned—the degradable residuals. But many residuals are non-degradable. These are not attacked by stream biota and undergo no great change once they get into a stream. In other words, the stream does not "purify itself" of them. This category includes inorganic substances—such materials as inorganic colloidal matter, ordinary salt, and the salts of numerous heavy metals. When these substances are present in fairly large quantities, they result in toxicity, unpleasant taste, hardness, and, especially when chlorides are present, in corrosion. These residuals can be a public health problem—usually when they enter into food chains. Two particularly vicious instances of poisoning by heavy metals have stirred the population of Japan. These are mercury poisoning through eating contaminated fish (Minimata disease) and cadmium poisoning through eating contaminated rice (Itai Itai disease). Several hundred people have been affected and more than a hundred have died. At the present time the Canadian government has forbidden the consumption of fish from both Lake Erie and Lake St. Clair because of feared mercury poisoning, and mercury has been discovered in many rivers in the United States.

Persistent pollutants

There is a third group of pollutants, mostly of relatively recent origin, which does not fit comfortably into either the degradable or non-degradable categories. These "persistent" or "exotic" pollutants are best exemplified by the synthetic organic chemicals produced in profusion by modern chemical industry. They enter watercourses as effluents from industrial operations and also as waste residuals from many household and agricultural uses. These substances are termed "persistent" because stream biota cannot effectively attack their complex molecular chains. Some degradation does take place, but usually so slowly that the persistents travel long stream distances, and in groundwater, in virtually unchanged form. Detergents (e.g., ABS), pesticides (e.g., DDT), and phenols (resulting from the distillation of petroleum and coal products) are among the most common of these pollutants. Fortunately, the recent development and successful manufacture of "soft" or degradable detergents has opened the way toward reduction or elimination of the problems associated with them, especially that of foaming. However, another problem associated with dry detergents has not been dealt with. These detergents contain phosphate "fillers" which may aggravate the nutrients problem.

Some of the persistent synthetic organics, like phenols and hard detergents, present primarily aesthetic problems. The phenols, for example, can cause an unpleasant taste in waters, especially when they are treated with chlorine to kill bacteria. Others are under suspicion as possible public health problems and are associated with periodic fish kills in streams. Some of the chemical insecticides are unbelievably toxic. The material endrin, which until recently was commonly used as an insecticide and rodenticide, is toxic to fish in minute concentrations. It has been calculated, for example, that 0.005 of a pound of endrin in three acres of water one foot deep is acutely toxic to fish.

Concentrations of the persistent organic substances have seldom if ever risen to levels in public water supplies high enough to present an *acute* danger to public health. The public health problem centers around the possible *chronic* effects of prolonged exposure to very low concentrations. Similarly, even in concentrations too low to be acutely poisonous to fish, these pollutants may have profound effects on stream ecology, especially through biological magnification in the food chain; higher creatures of other kinds—especially birds of prey—are now being seriously affected because persistent pesticides have entered their food chains. No solid evidence implicates present concentrations of organic chemicals in water supplies as a cause of health problems, but many experts are suspicious of them.

The long-lived radio-nuclides might also be included in the category of persistent pollutants. They are subject to degradation but at very low rates. Atomic power plants may be an increasingly important source of such pollutants. Generation of power by nuclear fission produces fission waste products which are contained in the fuel rods of reactors. In the course of time these fuels are separated by chemical processes to recover plutonium or to prevent waste products from "poisoning" the reactor and reducing its efficiency. Such atomic waste can impose huge external costs unless disposed of safely. A large volume of low-level waste resulting from the day-to-day operation of reactors can for the time being be diluted and discharged into streams, although the permissible standards for such discharge have recently been severely questioned in the United States, both outside and inside the Atomic Energy Commission.

"Hot" waste, containing long-lived substances such as radioactive strontium, cesium, and carbon, is in a different category from any other pollutant. So far, the only practical disposal method for high-level wastes is permanent storage. The "ultimate" solution to this contamination problem may be fusion energy which leaves no residuals except energy. But while some promising developments have occurred recently—especially in the Soviet Union—its development (if even possible) is, at least, decades away.

The range of alternatives

One of the most important features of the waterborne residuals problem, from the point of view of economic analysis, is the wide range of technical options which exist both for reducing the generation and discharge of wastes and for improving the assimilative capacity of watercourses. In industry, in addition to treatment, changes in the quality and type of inputs and outputs, the processes used, and by-product recovery are important ways of reducing residuals discharge. The capability of watercourses to assimilate residuals can often be

Table 1. *Summary of gaseous residuals from energy conversion, 1965 (million tons)*

Energy user	Carbon monoxide (CO)	Hydro- carbons (HC)	Sulfur dioxide (SO$_2$)	Oxides of nitrogen (NO$_x$)	Particu- lates
Utility power	1	neg.	13.6	3.7	2.4
Industry and households	5	neg.	8.4	7.0	7.0
Transportation	66	12	0.4	6.0	0.2
Total	72	12	22.4	16.7	9.6

neg.: Negligible.
Source: U.S. Public Health Service.

increased by using releases from reservoirs to regulate low river flows and by the direct introduction of air or oxygen into them by mechanical means.[1]

III. Airborne Residuals[2]

Types, sources, and management alternatives

There is virtually an infinity of airborne residuals that may be discharged to the atmosphere, but the ones of central interest and most commonly measured are carbon monoxide, hydrocarbons, sulfur dioxide, oxides of nitrogen, and particulates. The quantities and main sources of these in the United States are shown in the following table.

In the United States, by far the greatest tonnage of airborne residuals comes from the transportation sector, and virtually all of this is from internal combustion engines. They are especially important sources of carbon monoxide, hydrocarbons, and oxides of nitrogen. There are a number of ways in which emissions can be reduced from internal combustion engines, and some of these have been implemented. Carbon monoxide and hydrocarbons can be controlled to some extent by various means of achieving more complete combustion of the fuel delivered to the fuel tank. Oxides of nitrogen are much harder to deal with because they are not a result of incomplete combustion but are synthesized from atmospheric gases when combustion takes place under high temperatures and pressures. It is now thought that the best way to control these would be through catalytic afterburners which, however, could add substantially to the cost and complexity of engines. Many people

---

[1] A fairly extensive discussion of technical options can be found in A. V. Kneese and B. T. Bower, *Managing Water Quality: Economics, Technology, Institutions.* The Johns Hopkins Press, Baltimore, 1968.

[2] In preparing the section on air pollution, I have benefitted from an unpublished memorandum by Blair Bower and Derrick Sewell, 1969.

outside the automobile industry believe that large reductions in emissions can be effectively and economically achieved by abandoning the internal combustion engine in favor of other engine types (such as steam and electric) and heavier reliance on mass transit.[1]

Stationary sources (utility power, industry and households) are the main sources of sulfur oxides, particulates, and oxides of nitrogen. Control of emissions from these sources is a large and complex subject, but the main possibilities can be grouped into four categories: (1) fuel preparation (such as removing sulfur-bearing pyrites from coal before combustion), (2) fuel substitutions (such as substituting natural gas and low-sulfur oil and coal for high-sulfur coal), (3) redesigning burners (for example, in oil-burning furnaces two-stage combustion can reduce oxides of nitrogen), and (4) the treatment of stack gases (for example, stack gases can be scrubbed with water or dry removal processes can extract sulfur and particulates).[2]

Of course, all of these control technologies are likely to involve net costs even when they result in usable recovered materials. Furthermore, none of these processes inherently results in a reduction of CO_2. The possible significance of this was discussed in the opening section.

Assimilative capacity of the atmosphere

The capacity of the atmosphere to assimilate discharges of residuals varies with time, space, and the nature of the materials being discharged. From a resources management point of view it is necessary to be able to translate a specified time and location pattern of discharges of gaseous residuals into the resulting time and spatial pattern of ambient (environmental) concentrations, because in most cases there are multiple sources of discharge. With variations in type, quantity, and time pattern of discharge, the problem is compounded in complexity. However, imaginative applications of atmospheric diffusion models analogous to the water diffusion models described earlier have been used to help define "air sheds" for analysis of air quality management strategies.[3]

Another complication in environmental modeling results from the interactions between gaseous residuals and water quality. Such interactions can involve large geographic areas. For example, atmospheric scavenging— particularly washout by precipitation—appears to be becoming an increasing problem. High stacks are often used to reduce the local impact of air pollution, but they result in spreading the residual more thinly over larger areas.

---

[1] R. U. Ayres and R. P. McKenna: *Alternatives to the Internal Combustion Engine: Impacts on Environmental Quality*. The Johns Hopkins Press, Baltimore, 1971.

[2] A good source on the technology of pollution control from stationary sources is Arthur B. Stern (ed.), *Air Pollution*, vols. I, II, III. Academic Press, New York, 1968.

[3] A. A. Teller: The Use of Linear Programming to Estimate the Cost of Some Alternative Air Pollution Abatement Policies", *Proceedings* of the IBM Scientific Computing Symposium on Water and Air Resource Management, held on Oct. 23–25, 1967, at the Thomas J. Watson Research Center, Yorktown, Heights, N.Y.

Thus, high stacks on power plants in England are said to be causing "acid rain" in Scandinavia.

Precipitation is the primary cleansing mechanism for airborne gases and fine particles. Since sulfur dioxide is highly soluble in water, the washout process involves the absorption of the gas by drops of rain (or flakes of snow) as they fall through the gaseous discharge from a stack. Where the washout occurs over a body of water, adverse effects on water quality can occur. For example, atmospheric scavenging is believed to be contributing to deterioration of water quality in the Great Lakes Basin, at least with respect to the presence of trace elements in the lakes.

The areal extent of the atmospheric scavenging phenomenon is illustrated by data from the atmospheric chemical network stations in Europe relating to the acidity and sulfur contents of precipitation and the consequences on soils, surface waters, and biological systems. In 1958, pH values (pH is a measure of acidity—the lower the pH, the higher the acidity) below 5 were found only in limited areas over the Netherlands. In 1966, values below 5 were found in an area that spreads over Central Europe, and pH values in the Netherlands were less than 4.

Impacts of gaseous residuals on receptors

Perhaps of most immediate concern are direct effects on people, ranging in severity from the lethal to the merely annoying. Except for extreme air pollution episodes, fatalities are not, as a rule, traceable individually to the impact of air pollution, primarily because most of the effects are synergistic. Thus, air pollution is an environmental stress which, in conjunction with a number of other environmental stresses, tends to increase the incidence and seriousness of a variety of pulmonary diseases, including lung cancer, emphysema, tuberculosis, pneumonia, bronchitis, asthma, and even the common cold. Clearly, however, acute air pollution episodes have raised death rates. Such occurrences have been observed in Belgium, Britain, Mexico, and the United States, among others. But the more important health effects appear to be associated with persistent exposure to the degraded air which exists in most cities.

The preponderance of evidence suggests that the relationship between such pollutants as SO_2, CO, particulates, and heavy metals and disease is real and large.[1] But one should not underrate the difficulties of establishing such relationships in an absolutely firm manner.

Direct effects on humans have parallels in the animal and plant worlds. Animals of commercial importance (livestock) are not located to any appreciable extent within cities, so effects on them are usually minor. Effects on pets (dogs, cats, and birds) almost certainly exist, although they have not been much documented.

---

[1] L. B. Lave and E. P. Seskin: Air pollution and human health. *Science 169*, no. 3947 (Aug. 1970).

As far as plants are concerned, much the same situation holds. Crops are mostly some distance away from cities, and hazards are likely to be rather special in nature (e.g., fluorides from superphosphate plants, or sulfur oxides from copper smelters). However, there are some districts where truck crops—mostly fruits and vegetables—are grown in close juxtaposition to major cities and substantially affected by air pollution. In suburban gardens and city parks, there are deletrious effects on shrubs, flowers, shade trees, and even on forests in the air sheds of cities.

Damage to property

A third category of effects comprises damage to property. Here again, sulfur oxides and oxidants are perhaps equally potent. Sulfur oxides combine with water to form sulfurous acid (H_2SO_3) and the much more corrosive sulfuric acid (H_2SO_4). These acids will damage virtually any exposed metal surface and will react especially strongly with limestone or marble (calcium carbonate). Thus many historic buildings and objects (like "Cleopatra's Needle" in New York) have suffered extremely rapid deterioration in modern times.

Sulfur oxides will also cause discoloration, hardening and embrittlement of rubber, plastic, paper, and other materials. Oxidants such as ozone will also produce the latter type of effect. Of course, the most widespread and noticeable of all forms of property damage is simple dirt (soot). Airborne dirt affects clothing, furniture, carpets, drapes, exterior paintwork, and automobiles. It leads to extra washing, vacuum cleaning, dry-cleaning, and painting; and, of course, all of these activities do not entirely eliminate the dirt, so that people also must live in darker and dirtier surroundings.

A few comments comparing air and water pollution problems

There are important parallels and contrasts between the effects and possible modes of management of water and air pollution.

1. In the United States and abroad, air pollution is heavily implicated as a factor affecting public health. Water pollution may be more costly in terms of non-human resources, but the current link of water pollution to public health problems on any large scale in advanced countries is a matter of suspicion concerning chronic effects rather than of firm evidence. Much stronger evidence links air pollution to public health problems.

2. As in the case of water pollution, a great many of the external costs imposed by air pollutants would appear to be measurable, but very little systematic measurement has yet been undertaken. The more straightforward effects are, for example, soiling, corrosion, reduction in property values, and agricultural losses.[1]

3. Current technology apparently provides fewer classes of means of dealing

---

[1] Some efforts to provide economically useful estimates of damages are discussed elsewhere in this volume.

with air pollution than with water pollution. In part this is because it is easier for man to control hydrological events than meteorological events. The assimilative capacity of the air mantle cannot be effectively augmented. In part it is because air is not delivered to users in pipes as water frequently is, so that it is only to a limited extent that polluted air is treatable before it is consumed. Therefore, we are in somewhat the same position in regard to polluted air as the fish are with polluted water. We live in it. Furthermore, it is also more difficult and costly to collect gaseous residuals for central treatment. Accordingly, control of air pollution is largely a matter of preventing pollutants from escaping from their source, eliminating the source, or of shifting location of the source or the recipient. Water pollution, on the other hand, is in general subject to a larger array of control measures. Nevertheless, both present intricate problems of devising optimal control systems.

4. To the extent that air sheds are definable, air shed authorities or compacts of districts are conceivable and may be useful administrative devices. In the United States the current federal policy approach points strongly in this direction and in this respect (but not others) is more advanced than the water pollution control programs.

IV. Solid Residuals

Just about every type of object made and used by man can and does eventually become a solid residual. Some of the main categories of importance are organic material, which includes garbage, and industrial solid wastes, such as from the canning industry, for example. Newspapers, wrappings, containers, and a great variety of other objects are found in household, commercial, office, industrial solid wastes. A very important source of solids in the United States is automobiles which will be discussed separately, further on. In the United States, about 5 lb. per day of solid wastes are collected of which about three are household and commercial. Industrial, demolition, and agricultural wastes constitute most of the other. Altogether the United States generates (exclusive of agricultural wastes) each year approximately 3–5 billions tons of solid residuals from household, commercial, animal, industrial, and mining activities and spends about 4–5 billion dollars to handle and dispose of them.[1] In addition, there is a large amount of uncollected solids which litter the countryside.

The disposal of solid wastes can have a number of deleterious effects on society. Littering, dumps, and landfills produce visual disamenities. The disposal of solid wastes can cause adverse effects on air and water quality. Incineration of solid wastes is an obvious source of air pollution in many areas, as are burn-

---

[1] R. Black, A. Muhich, A. Klee, H. Hickman and R. Vaughn: *The National Solid Waste Survey.* U.S. Department of Health, Welfare, and Education, 24 Oct. 1968. Also W. O. Spofford, Jr. Solid waste management: Some economic considerations. *Natural Resources Journal,* vol. 11 (forthcoming: 1971).

ing dumps. Dumps tend to catch fire by spontaneous combustion unless they are relatively carefully controlled. Furthermore, drainage from disposal sites can reduce water quality in watercourses, and the sites may also provide a habitat for rodents and insect vectors. The disposal of collected solid wastes (which excludes much industrial waste, automobiles, and all of agricultural waste) in the United States is roughly in the following proportions: about 90 % goes into landfill operations of one kind or another; another 8 % is incinerated; and a small amount, about 2 %, goes into hog-feeding and miscellaneous categories. Landfill can be an effective and low external cost way of disposing of wastes. However, many landfills are poorly operated and impose external costs via effects on the air, water, and landscapes.

As previously mentioned, automobiles are a special problem. Of the 10–20 million junk cars in existence at any one time in the United States, about 73 % are in the hands of wreckers, in other words, in junk yards; about 6 % in the hands of scrap processers; and about 21 % abandoned and littering the countryside. Recovery could be made much more economical by slight design changes, but presently there are no incentives to do so. Furthermore, unless it is managed to avoid them, the recycle or "secondary materials" industry itself can cause substantial external costs. For example, automobiles are usually burned prior to being prepared for scrap metal, and this can be an important source of local air pollution. Some of the processes involved in recycling automobiles have very high noise levels. While it seems clear that recycle of materials is underused under present circumstances, suffering as it does from tax, labelling, and other disadvantages with respect to new materials, it is also true that it is not a total panacea as one might gather from some of its more ecstatic adherents. For example, the paint on automobiles could not be recycled except at the expense of immense quantities of energy and other resources. Moreover, some materials such as paints, thinners, solvents, cleaners, fuels, etc., cannot perform their functions without being dissipated to the environment.

V. The Flow of Materials[1]

To tie together some of the points made in previous sections, it is useful to view environmental pollution and its control as a materials balance problem for the entire economy. Energy residuals could be treated in an entirely parallel fashion, but I will not discuss this here.[2]

The inputs to the system are fuels, foods, and raw materials which are

---

[1] This section is based heavily on R. U. Ayres and A. V. Kneese: "Production, consumption, and externalities." *American Economic Review 59*, no. 3 (June 1969).

[2] While very little direct exchange between material and energy occurs, it is important to note that there are significant tradeoffs between these residuals streams. For example, an effort to achieve complete recycle with present levels of materials flow would require monstrous amounts of energy to overcome entropy.

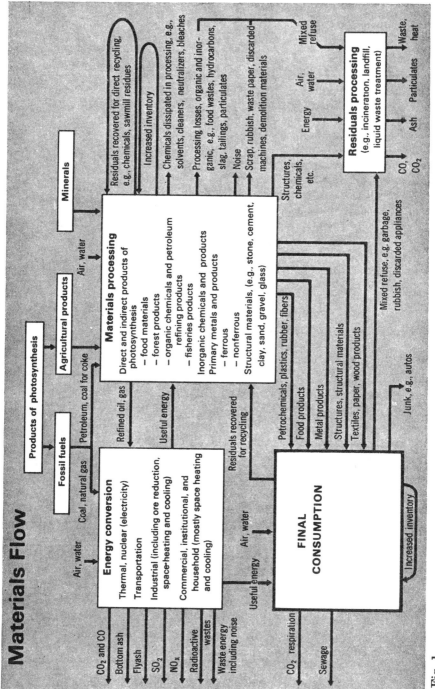

Fig. 1.

partly converted into final goods and partly become residuals. Except for increases in inventory, final goods also ultimately enter the residuals stream. Thus goods which are "consumed" really only render services temporarily. Their material substance remains in existence and must either be reused or discharged to the environment.

In an economy which is closed (no imports or exports) and where there is no net accumulation of stocks (plant, equipment, inventories, consumer durables, or residential buildings), the amount of residuals inserted into the natural environment must be approximately equal to the weight of basic fuels, food, and raw materials entering the processing and production system, plus gases taken from the atmosphere. This result, while obvious upon reflection, leads to the at first rather surprising corollary that residuals disposal involves a greater tonnage of material than basic materials processing, although many of the residuals, being gaseous, require no physical "handling."

Fig. 1 shows a materials flow of the type I have in mind in a little greater detail and relates it to a broad classification of economic sectors. In an open (regional or national) economy, it would be necessary to add flows representing imports and exports. In an economy undergoing stock or capital accumulation, the production of residuals in any given year would be less by that amount than the basic inputs. In the entire U.S. economy, accumulation accounts for about 10–15 % of basic annual inputs, mostly in the form of construction materials, and there is some net importation of raw and partially processed materials amounting to 4 or 5 % of domestic production. Table 2 shows estimates of the weight of raw materials produced in the United States in several recent years, plus net imports of raw and partially processed materials.

Of the active inputs,[1] perhaps three-quarters of the overall weight is eventually discharged to the atmosphere as carbon (combined with atmospheric oxygen in the form of CO or CO_2) and hydrogen (combined with atmospheric oxygen as H_2O) under current conditions. This results from combustion of fossil fuels and from animal respiration. Discharge of carbon dioxide can be considered harmless in the short run, as we have seen, but may produce adverse climatic effects in the long run.

The remaining residuals are either gases (like carbon monoxide, nitrogen dioxide, and sulfur dioxide—all potentially harmful even in the short run), dry solids (like rubbish and scrap), or wet solids (like garbage, sewage, and industrial wastes suspended or dissolved in water). In a sense, the dry solids are an irreducible, limiting form of waste. By the application of appropriate equipment and energy, most undesirable substances can, in principle, be removed from water and air streams[2]—but what is left must be disposed of in solid form, transformed, or reused. Looking at the matter in this way

---

[1] See footnote to Table 2.
[2] Except CO_2, which may be harmful in the long run, as noted.

Table 2. *Weight of Basic Materials Production in the United States, plus Net Imports, 1963–65 (million tons)*

	1963	1964	1965
Agricultural (incl. fishery and wildlife and forest) products			
Food and Fiber			
Crops	350	358	364
Livestock and diary	23	24	23.5
Fishery	2	2	2
Forestry Products			
Forestry Products			
(85% dry wt. basis)			
Sawlogs	107	116	120
Pulpwood	53	55	56
Other	41	41	42
Total	576	596	607.5
Mineral Fuels			
Total	1 337	1 399	1 448
Other Minerals			
Iron ore	204	237	245
Other metal ores	161	171	191
Other non-metals	125	133	149
Total	490	541	585
Grand Total[a]	2 403	2 536	2 640.5

[a] Excluding construction materials, stone, sand, gravel, and other minerals used for structural purposes, ballast, fillers, insulation, etc. Gangue and mine tailings are also excluded from this total. These materials account for enormous tonnages but undergo essentially no chemical change. Hence, their use is more or less tantamount to physically moving them from one location to another. If these were to be included, there is no logical reason to exclude material shifted in highway cut-and-fill operations, harbor dredging, landfill plowing, and even silt moved by rivers. Since a line must be drawn somewhere, we chose to draw it as indicated above.

Source: R. U. Ayres and A. V. Kneese: Environmental Pollution. In *Federal Programs for the Development of Human Resources*, a Compendium of Papers submitted to the Subcommittee on Economic Progress of the Joint Economic Committee, United States Congress, Vol. 2. Government Printing Office, Washington, 1968. Some revisions have been made in the original table.

clearly reveals a primary interdependence among the various residuals streams which casts into doubt the traditional classification, which I have used earlier in this article, of air, water, and land pollution as individual categories for purposes of planning and control policy.

Residuals do not necessarily have to be discharged to the environment. In many instances, it is possible to recycle them back into the productive system. The materials balance view underlines the fact that the throughput of new materials necessary to maintain a given level of production and consumption decreases as the technical efficiency of energy conversion and materials utilization and reutilization increases. Similarly, other things being equal, the longer cars, buildings, machinery, and other durables remain in service, the fewer new materials are required to compensate for loss, wear, and obsolescence—although the use of old or worn machinery (e.g., automobiles) tends to increase other residuals problems. Technically efficient combustion of (desulfurized) fossil fuels would leave only water, ash, and carbon dioxide as residuals, while nuclear energy conversion need leave only negligible quantities of material residuals (although thermal pollution and radiation hazards cannot be dismissed by any means).

Given the population, industrial production, and transport service in an economy (a regional rather than a national economy would normally be the relevant unit), it is possible to visualize combinations of social policy which could lead to quite different relative burdens placed on the various residuals-receiving environmental media; or, given the possibilities for recycle and less residual-generating production processes, the overall burden to be placed upon the environment as a whole. To take one extreme, a region which went in heavily for electric space heating, electric transportation systems, and wet-scrubbing of stack gases (from steam plants and industries), which ground up its garbage and delivered it to the sewers and then discharged the raw sewage to watercourses, would protect its air resources to an exceptional degree. But this would come at the sacrifice of placing a heavy residuals load upon water resources. On the other hand, a region which treated municipal and industrial waste water streams to a high level and relied heavily on the incineration of sludges and solid wastes would protect its water and land resources at the expense of discharging waste residuals predominantly to the air. Finally, a region which practiced high-level recovery and recycle of waste materials and fostered low residual production processes to a far-reaching extent in each of the economic sectors might discharge very little residual waste to any of the environmental media.

Further complexities are added by the fact that sometimes it is, as we have seen, possible to modify an environmental medium through investment in control facilities so as to improve its assimilative capacity. The easiest to see but far from only example is with respect to watercourses where reservoir storage can be used to augment low river flows that ordinarily are associated with critical pollution (high external cost situations). Thus, internalization of external costs associated with particular discharges, by means of taxes or other restrictions, even if done perfectly, cannot guarantee Pareto optimality. Collective investments involving public good aspects must enter into an optimal solution.

To recapitulate the main points these considerations raise for economic analysis briefly: (1) Technological external diseconomies are not freakish anomalies in the processes of production and consumption but an inherent and normal part of them. Residuals generation is inherent in virtually all production and consumption activities, and there are only two ways of handling them—recycle, or discharge into environmental media without or with modifications. (2) These external diseconomies are apt to be quantitatively negligible in a low-population or economically undeveloped setting, but they become progressively (nonlinearly) more important as the population rises and the level of output increases (i.e., as the natural reservoirs providing dilution and other assimilative properties become exhausted). (3) They cannot be properly dealt with by considering environmental media, such as air and water, in isolation. (4) Isolated and ad hoc taxes and other restrictions are not sufficient for their optimum control, although taxes and restrictions are essential elements in a more systematic and coherent program of environmental quality management. (5) Public investment programs, particularly including transportation systems, sewage disposal systems, and river flow regulation, are intimately related to the amounts and effects of residuals and must be planned in light of them. (6) There is a wide range of (technological) alternatives for coping with the environmental pollution problems stemming from liquid, gaseous, and solid residuals. Economic tools need to be selected and developed which can be used to approximate optimal combinations of these alternatives.

Part II
THEORY AND POLICY

ESSAY ON ECONOMIC GROWTH AND ENVIRONMENTAL QUALITY*

*Ralph C. d'Arge***

University of California, Riverside, California, USA

Summary

The major premise of this essay is that production-consumption and waste emissions tend to be joint products of the human species. Given this premise, a parable of an astronaut irretrievably lost in space is discussed in order to deduce several propositions on the astronaut's optimal rate of consumption. A Harrod type of model is also analyzed with regard to the rate of consumption over time where the model includes a simplified depiction of the interaction between the economy and natural environment. Empirical estimates of the impact of effluent charges on comparative international advantage of selected countries are also presented. The major conclusion is that national and international economic policies and national environmental policies are not separable.

I

Pervasive environmental changes accompanying the development of modern societies have wrought doubts regarding the insignificance of uncontrolled interdependencies between man and the natural environment. The economic acts of production and consumption not only benefit man through his provision but simultaneously involve the creation of wastes with the potential of degrading his environment. It can be asserted that production-consumption and waste emission are joint products of the human species ... they certainly are for all other animal species and plant life. A conservative estimate of waste flow in the United States currently would be 6–7 pounds of "active" waste produced per dollar of gross national product, and there is some evidence that waste flow increases at a faster rate than income.[1] This would

* A portion of the research reported on here was financially supported by Resources For The Future, Inc., Washington, D.C. with no responsibility for the inferences or results contained herein.
** I wish to acknowledge the comments and criticisms of the following individuals without committing them to agreement with any part of this essay: K. C. Kogiku, T. Crocker, K. Oddson, E. Brook, O. Bubik, T. Clark, and H. Lawton.
[1] See R. U. Ayres and A. V. Kneese, "Environmental Pollution" in United States Congress, Joint Economic Committee, *Federal Programs for the Development of Human Resources*, Vol. 2, Washington, D. C. (1968); P. Leighton, "The Air Resource", in proceedings of Statewide Conference on: *Man in California—1980's*, Sacramento, California (January 1964); or T. Scitovsky, "External Diseconomies in the Modern Economy", *Western Economic Journal*, Vol. IV (summer 1966).

undoubtedly be true for less developed economies attempting to industrialize with a very small services sector. In highly developed economies the inter-dependencies between the economy and natural environment are accentuated because of vast agglomerations of people and high waste generating industries in locations with only limited waste assimilative capability. The natural environment containing highly industrialized and densely grouped populations can no longer be viewed as a bathtub by economists, where once each month the ring needs to be cleaned, but the drain never clogs. As an example, the somewhat isolated and rare disasters in the Meuse Valley, Donora, Pennsylvania, and London due to "smogs" are rapidly becoming seasonal threats in such cities as Los Angeles and Tokyo.

The purpose of this essay is to analyze some of the basic issues involved in the dichotomy between economic growth and environmental degradation and conceivably, decimation. We shall start by discussing an extremely simple parable about an economic spaceman and then attempt to relate it to global decisions with respect to the environment. I believe several enlightening propositions can be obtained from such an extreme parable. Next, a brief exposition of the interconnection between macro growth models and the natural environment will be made, again to attempt to find some useful propositions. Our third task will be to discuss briefly and present tentative empirical evidence on some of the major difficulties in developing international agreements for the regulation of waste flows.

Before presenting the spaceman parable, a comment is deserved on the connections between highly localized and interregional-international problems of environmental pollution. One factor, in my opinion, which has led economists to ignore the general interdependency questions of man-environment themes has been the apparent implicit belief that externalities associated with pollution were essentially two-party problems ... the emittor versus the receptor.[1] If emittor and receptor are identified, Coase, Buchanan, Turvey, and others have pointed out that through private exchange, regardless of who owns the pollutant receiving medium, efficiency will result through the equating of marginal pollutant damages and marginal waste treatment costs.[2] These cases, if they existed at all, have already been resolved under the aegis of private property rights and legal adversary proceedings, at least in the western countries. However, the less mundane but pervasive cases where receptors are only vaguely identified and (or) damages induced by emittors are synergistic in their effects on receptors cannot be resolved with the application of two-party principles of property rights.

[1] For an extremely lucid discussion of the problems encountered utilizing the "two-party" mentality, see A. V. Kneese, "Environmental Pollution and Economic Policy", to be presented to the American Economic Association, Detroit, Michigan (December, 1970).
[2] See the definitive articles by: R. H. Coase, "The Problem of Social Cost", *Journal of Law and Economics* (October, 1960), J. M. Buchanan and W. C. Stubblebine, "Externality", *Economica* (November, 1962), and R. Turvey, "On Divergencies Between Social Cost and Private Cost", *Economica* (August, 1963).

As Allen Kneese has frequently argued, the receptor in most pervasive cases of environmental pollution is or cannot be well defined ... usually being a group of people, both present and future generations, without a viable organization to represent their interests.[1] Recipients of the effects of long-term accumulations of DDT, mercury, or CO_2 in the atmosphere have little or no voice in the decision on emission rates of these potentially dangerous pollutants with nearly global dispersion. River basin commissions such as are found in the Ruhr industrial basin or for the Delaware in the United States may improvise sub-optimal plans for maintaining water quality within a basin, but how cognizant are these commissions of pollutant loads at the mouth of the river? Relatively high water quality standards or effluent charges will undoubtedly induce shifts in production processes toward a greater pollutant load to the atmosphere via burning or industrial sludge, or toward creating solid waste disposal problems, unless the air and land resources are also protected by standards or effluent charges.

It seems to me that in concentrating on two-party or localized effects, economists have not taken heed of the warnings of ecologists and have ignored one of the major dictums of economic thought: "that everything depends on everything else".[2] In theoretical discussions of environmental control, I would take the side of the ecologists who not only echo the general equilibrium principle of economics, but demand that it be applied in analyzing issues of environmental pollution. A general equilibrium stance should be adopted which attempts to recognize at least the major interdependencies between man's economic behavior and the effect and interaction of such behavior on the environment in which he resides. What follows must be viewed in this light even though at most the concepts and parables discussed are no more than suggestive of extremely feeble beginnings.

---

[1] A. V. Kneese and B. T. Bower, *Managing Water Quality: Economics, Technology, and Institutions*, Baltimore: Johns Hopkins Press (1968).

[2] In several recent instances, economists have attempted to integrate some rather simple concepts of the man-environment interaction. For an integration of economic equilibrium and materials balance, see R. U. Ayres and A. V. Kneese, "Production, Consumption, and Externalities", *American Economic Review* (June, 1969); and for a less rigorous but philosophical categorization on a similar theme, H. Daley, "Economics as a Life Science", *Journal of Political Economy* (April, 1968). For a discussion of Pareto optimality, second best, and environmental taxation within the context of materials balance, see A. V. Kneese and R. C. d'Arge, "Pervasive External Costs and the Response of Society", United States Congress, Joint Economic Committee, *Economic Analysis of the P.P.B.S. System ...*, Washington, D.C. (May, 1969). On production and renewable resources, see V. L. Smith, "The Economics of Production from Natural Resources", *American Economic Review*, Vol. LVIII (June, 1968), and R. G. Cummings and O. R. Burt, "The Economics of Production from Natural Resources: Note", *American Economic Review*, Vol. LIX (December, 1969). For an integration of pollution and utilization of exhaustible resources concepts, see R. C. d'Arge, "Economic Growth, Recycling, and the Natural Environment", in A. V. Kneese, editor, *Environmental Quality Analysis: Research in the Social Sciences*, Baltimore: Johns Hopkins Press (forthcoming, 1970).

II

The analogy applied here is with spacemen since astronauts have replaced cowboys and Vikings as recipients and victims of the laurels of contemporary hero worship. What is deduced in the following paragraphs is equally applicable in many instances to the former hero's "rational" decisions to "break camp" or depart from Jutland.

Let us assume that there is a lone astronaut within a sealed space capsule who is suddenly confronted with the realization that he will not be able to return to earth. He is irretrievably lost in space. Within his capsule are stored enough provisions to sustain life for a period of time shorter than the astronaut's "normal life" span even if he consumes at minimal subsistence in terms of caloric intake. Clearly if the spaceman's criterion of welfare is the singular objective of maximizing the length of existence, he will not consume above the minimum subsistence level. Alternatively, he may be more or less myopic, depending on one's philosophy, and decide to consume at a faster rate, particularly if he evaluates an additional year of existence at near starvation levels to be of less value than an increment of present consumption. The rational spaceman would consume at a rate during each period of consumption such that he equated the marginal utility of a unit of consumption in that period with the foregone opportunity, in terms of utility, of consuming that unit in any future consumption period including the period following the one in which he runs out of provisions. The price of this foregone opportunity will also include a positive shadow price for not being able to consume below a minimal subsistence level if such a level is constraining to the astronaut's desired rate of consumption. These consequences are obvious and are related to a negative analogue of Ramsey's theory of saving.[1] However, here the astronaut dissaves and approaches a finite anti-bliss point. The conclusions stipulated are also nearly identical to recent models of mine exploitation.[2]

From this extremely naive model we are able to make several additional inferences. First, if the astronaut has a positive finite rate of time preference he will consider consuming at a higher rate than minimal subsistence only if his evaluation of an additional time interval of survival is not infinite. Second, if his preferences and capacity are such that he exhibits diminishing marginal utility of consumption, he is likely to spread consumption over a longer time span than if he achieved bliss through an immediate orgy of satiation.

Next, let us revise our assumptions and introduce the idea that since the astronaut resides in a sealed capsule he must endure the effects of his own waste generation. The astronaut is now confronted with the problem that higher immediate rates of consumption will mean a higher density of wastes

[1] F. Ramsey, "A Mathematical Theory of Saving", *Economic Journal*, Vol. 38 (1928).
[2] See R. G. Cummings and O. K. Burt, op. cit.

to live with during future time intervals. In order to analyze the question of rational planning by the astronaut, let us stipulate the following assumptions: (*a*) The astronaut's preferences are such that he has diminishing marginal utility of consumption and increasing marginal disutility induced by expanding waste density; (*b*) There is some waste density or concentration which is lethal to the astronaut and he is aware of this; (*c*) There is very little or no assimilation of wastes or recycling within the sealed capsule; (*d*) The astronaut has enough provisions to last his normal lifetime and this lifetime is a very large number such that his only binding constraint is his generation of wastes; (*e*) Consumption and waste emissions are joint products, i.e., the principles of conservation of matter-energy is operative; and (*f*) The astronaut is not decreasing in size and does not discount future consumption at a negative or too high a positive rate. Given these assumptions, the astronaut following a rational policy of maximizing utility over time will follow one of two rules:

Proposition 1. If maximizing the length of existence is of primary concern, the astronaut will immediately reduce consumption to the minimal subsistence level and remain there until decimation occurs.
Proposition 2. If continued existence is not of overriding importance, the astronaut will immediately reduce consumption upon discovering that he is lost in space but increase it thereafter.

The formal proof of Proposition 2 is given elsewhere[1] and the logic of Proposition 1 is clearly seen from the earlier discussion of the astronaut's dilemma on dissaving over time. Clearly, the mining and dissaving problem is identical to the pollution problem in many ways but different in one vital respect ... dissaving to the astronaut *per se* does not cause disutility whereas waste generation is usually assumed to.

Professor Robert Strotz came up with what I believe to be an ingenious and simple explanation for Proposition 2. He stated the problem as one of campers entering a campground and planning to stay for some finite interval of time. The campers are confronted with the problem that if they consume their provisions and discard waste at a high rate initially they will have to suffer the "disutilities" of this waste during the remainder of their stay. However, if the campers consume at a low rate initially and build up consumption over time, they need not suffer too much from waste generated in early intervals and yet increasing consumption allows them to compensate for the increasing waste density. At some point they will depart leaving a "debacle of debris" following a "grand orgy of consumption".

Like the factor price equalization theorem of international trade, these propositions at best have very tenuous connections with reality. However,

[1] See R. C. d'Arge, "Economic Growth, Recycling ...", op. cit. or R. C. d'Arge and K. C. Kogiku, "Economic Growth and the Natural Environment", paper to be given at the Econometric Society Meetings, Detroit, Michigan (December, 1970).

I believe these propositions lead to some important thoughts on very long-run environmental quality problems. If the earth is truly a spaceship or closed resource system characterized by finite amounts of energy and natural waste assimilative capability, energy dispersion coupled with environmental degradation will, at some finite point in time, severly limit man's proclivities for reproduction and greater material abundance. And even if the finite planning interval is assumed to be extremely large, say the length of time before the sun's gases undergo complete combustion, we are confronted with the fact that optimization, under a set of environmental constraints, generally leads to minimal subsistence or at least, retarded current consumption.[1] This is particularly true if population is expected to continue growing at positive rates for the indefinite future. Not only is the environmental quality management problem one of applying effluent charges, standards, and (or) payments which reflect current damages to the population, but these charges should also include a substantial adjustment for the effects on future generations, especially for reducing opportunities of a higher standard of living which includes less environmental degradation. Thus, it appears that environmental quality management must involve the analysis of what might be termed dynamic or time related externalities ... the impact of one generation's ignoring discrepancies between private and social costs on subsequent generations' economic choices. Perhaps there is an emerging rationale in these simple propositions for placing a ceiling on per capita consumption or even considering the possibility of zero growth of material goods in more advanced economies.

The spaceship analogies discussed here contain several major weaknesses. Always implicit within the previous discussion of spaceships are the very long-run facets of survival conditioned by limited energy reserves, energy dispersion, an expanding or constant population, and no compensating investments in recycling technologies or in augmenting the natural environment's capacity to assimilate wastes. Of course, these assumptions yield the most cataclysmic outcomes. Next, we turn to a discussion within the context of

---

[1] It should be noted that a finite planning interval coupled with a finite environmental capacity for wastes are extremely rigid assumptions. Of course, infinite planning intervals in conjunction with an unlimited natural environmental assimilative capacity reduce to trivial nonexistent scarcity cases. Alternatively, an infinite planning interval in conjunction with a finite environmental waste assimilative capability appears to be philosophically inconsistent provided population is increasing over time, consumption per capita has a positive lower bound, rates of environmental waste assimilation are very low or zero, waste generating technology is completely static in character, and wastes at some level of concentration become lethal to the human species. These conditions quite obviously insure that at some moment in the finite future decimination will occur. The alternative case of a finite planning interval and infinite waste assimilative capability is also philosophically inconsistent. Specifying an infinite planning interval for mankind, of course, implies the imposition of a constraint on the decision process, namely that regardless of how degraded man's habitat becomes, he must continue to exist. A more general and encompassing decision process would stipulate that the survival period be optimally chosen with infinity as one possible choice. I have chosen the course of assuming a finite planning interval, thinking that finite numbers can be very large. However, such a decision precludes consideration of an infinite time horizon.

neo-classical economic growth in which the outcomes are not so bleak by introducing the possibilities of recycling investments and investments which increase the natural environment's capability to alter, assimilate, or otherwise neutralize wastes.

III

Saving or the augmentation of the existing capital stock plays a key role in contemporary thought on economic growth. In the simple Harrod types of growth, the greater the proportion of current output saved and invested the faster the economy is warranted to grow. For economies with the option to invest in augmenting waste assimilative capability through such investments as reservoir construction to provide stabilized river flows, waste treatment plants, or catalytic mufflers on internal combustion engines, saving becomes the source of funds for such investments. But saving also has an additional effect on the environment in that the higher the saving rate in relation to depreciation or decay rates the smaller the current flow of wastes. Thus, saving can have the impact of redistributing waste flows or residuals accruing to the environment over more extended intervals of time.

Investment directed toward reducing environmental degradation appears to be able to take three rather general forms: recycling, augmentation of the natural environment's assimilative capability, or investments which reduce the need for, or can compensate for, increased consumption of material goods. Recycling investments reduce waste loads by channeling wastes of the production and consumption processes back into these processes. Investments in augmenting assimilative capacity reduce the effects of waste loads by altering or augmenting the natural environment through such activities as reservoir construction, stream reaeration, and chemical and biological waste treatment. The third category of investments are directed toward inducing the populace toward consuming less time devoted to material consumption and more time toward services, solitude, or recreational-aesthetic pursuits involving little or no waste generating goods. In the model which follows, we will concentrate only on one category of such investments, namely those which augment environmental assimilative capability and (or) the special category of recycling investments which do not increase or decrease productive output.

In order to describe briefly some of the ramifications of growth and environmental quality, we shall adopt an extremely simple model of the Harrod type. Let W denote a homogeneous waste flow, F the current production of material goods, K_r a measure of the capital stock committed to augmenting the assimilative capability of the environment with $dK_r/dt = I_r$, and S saving. Then, in the spirit of the concept that production-consumption and waste generation are joint products, the following equation is postulated:

$$W = g_c(F - S) + g_f F \qquad (1`$$

where g_c and g_f are coefficients relating the amount of waste flow generated from current consumption and production, respectively. Next, it is postulated that density of wastes is the proper measure of environmental quality. Let D denote average waste density, V a volume measure of the natural environment, for example the size of the global natural life zone which is assumed fixed, $\dot{D}$ definitionally equal to dD/dt, and δ a decay rate in waste density which can be considered as symbolic of the natural assimilative capability of the environment:

$$\dot{D} = \frac{1}{V} W - hI_r - \delta \qquad (2)$$

where h is a coefficient reflecting the rate of augmentation of the natural assimilative capability of the environment by capital investments as previously described.[1] For now, we shall assume K_r does not depreciate though the model could easily be made to accommodate instantaneous and constant decay rates in K_r. From equations (1) and (2) we obtain a simple expression relating changes in waste density to investment in assimilative capability, current saving and output.

$$\dot{D} = \frac{1}{V} (g_c + g_f) F - \frac{1}{V} g_c S - hI_r - \delta \qquad (3)$$

Note from equation (3) that even if current saving and investment in augmenting the environment's assimilative capacity are both zero, there is a positive rate of output which would yield no increase or decrease in waste density. In ecologist's terms, this is a natural rate of production which hypothetically at least, balances man's proclivity to produce waste with nature's ability to absorb it ... a biological equilibrium.

To complete this simple model, define K_f as the non-depreciating capital stock devoted to producing material output, s as the marginal and average propensity to save from output, and σ as the incremental output–capital ratio. Then:

$$sF = S = I_f + I_r \qquad (4)$$

$$\dot{F} = \sigma I_f \qquad (5)$$

with $dK_f/dt = I_f$, and $dF/dt = \dot{F}$. Substituting equation (5) into (4), the connection between this and the Harrod-Domar model is observed:

$$\dot{F}/F = G_f = s\sigma - \sigma(I_r/F) \qquad (6)$$

Thus, if investment in non-productive waste assimilation is not undertaken, one ends up with a warranted rate of growth (G_f) equivalent to $s\sigma$. Such a

---

[1] Specifying h to be a constant in equation (2) carries with it the implicit assumption that the *level* of waste density at any point in time is linearly determined by past levels of production, assimilative investment and the cumulative neutralizing effects of the natural environment on waste density.

rate would be warranted if the natural environment could be characterized as limitless in capacity or size and in regenerative capability. Substituting (6) and (4) into (3) with rearrangement of terms, one obtains:

$$\left[\frac{g_c}{V}(1-s)+\frac{g_f}{V}-hs\right]F+\frac{h}{\sigma}\dot{F}-\delta=\dot{D} \tag{7}$$

In order to obtain a warranted rate of growth, the constraint is postulated of no change in average waste density, or $\dot{D}=0$. Such a constraint might not be desirable in that variations in waste density over time may yield greater welfare and (or) less disutility associated with waste flows. Such variations have been briefly mentioned earlier in conjunction with the space-man concept. However, in terms of the model proposed here it seems preferable to presume a "steady state" in terms of environmental degradation via waste flows. Such a steady state could proceed without the hindrance of a finite planning interval for man's survival, or the philosophical contradiction between finite resources and an infinite planning horizon. It should be pointed out, however, that the following references to a warranted rate of growth presume not only a non-deteriorating natural environment but also one that is not allowed to improve either. With the assumption $\dot{D}=0$, equation (7) becomes a first order non-homogeneous linear differential equation of the form:

$$\dot{F}+\alpha F-\eta=0 \tag{8}$$

and

$$\alpha=\sigma\left[\frac{g_c}{hV}(1-s)+\frac{g_f}{hV}-s\right]$$

$$\eta=\frac{\sigma\delta}{h}$$

which has a solution:

$$F(t)=He^{-\alpha t}+\eta/\alpha \tag{9}$$

where H is a constant determined by initial conditions. Given the assumption that $H>0$, for output to increase over time, $\alpha<0$ or $sh>g_c/V(1-s)+g_f/V$. This simply indicates saving rates and the efficiency of investment to reduce waste density must be high enough to compensate for the wastes generated through consumption and production.[1] As would be expected, the greater the

---

[1] Take $F(0)=F_0>0$; then at $t=0$, $H=F_0-\eta/\alpha$ and therefore:
$F(t)=F_0e^{-\alpha t}+\eta/\alpha(1-e^{-\alpha t})$
Thus, $\dot{F}>0$, provided $\alpha<0$, or $\alpha>0$ and $F_0<\eta/\alpha$. In words, if initial output is less than the "biological equilibrium", production is warranted to increase even though the economy cannot sustain $\dot{D}=0$ above the "biological equilibrium" without governmental controls. In this discussion we shall presume $F_0>\eta/\alpha$.

savings rate, efficiency of investment in the natural environment's assimilative capability, size or extent of the natural environment, and the smaller the amount of waste generated per unit of output and consumption, the more likely a sustainable positive rate of growth would be which would not simultaneously involve an increasing waste density. The incremental output-capital ratio alternatively is neutral, either accelerating the rate of growth in output if the warranted rate is positive or increasing the negative change in output toward the biological equilibrium.

In order to give some empirical content to the meaning of a warranted positive or negative change in output, let us redefine the h coefficient, which connects reduction in waste density to assimilative investment, in terms of waste magnitude, i.e., $h' = h \cdot V$ where h' denotes the reduction in waste tonnage polluting the environment per dollar of assimilative investment. Next, let the savings rate equal 0.20 and the sum of g_c and g_f equal 6 pounds. The latter is consistent with the estimate for the United States reported earlier. Then the criterion for a positive rate of growth in output is $h' > 30\text{-}g_c$. In other words, if pounds of waste generated per dollar of consumption expenditure equal two, then an average dollar of investment in waste assimilative capacity must yield at minimum a reduction of twenty-eight pounds in terms of waste density. The efficiency of waste assimilative investments must be substantially greater for lower savings rates. Lowering the rate to .10 requires investments in waste assimilation via augmenting the natural environment to rise to fifty-eight pounds per dollar of investment.

This model is perhaps the most simple that can be constructed which includes the major basic elements of both a growing economy and waste generation-assimilation processes. It embodies many extreme simplifications which undoubtedly will not withstand the probing of empirical verification.[1] In its defense, one can indicate that waste flows and increasing waste density relationships are not tied to any specific environmental phenomenon. The model is equally applicable to problems of mercury, lead, DDT, or BOD concentrations where assimilative investment may be chemical or physical processes of neutralization, constructing reservoirs to increase river flows during seasonal water shortages, or mechanical stream reaeration which raises

---

[1] The constant incremental output–capital ratio can be generalized by including a constant returns to scale production function, with the usual restriction on convexity with labor services as an added argument. In addition, labor force can be assumed to be a constant proportion of population, and a population growth equation with constant percentage growth rate can be included. These complications add the additional restriction that population cannot be increasing at too rapid a rate if output per capita is to be increasing and with $\dot{D} = 0$. But a compensating force arises in that labor can be substituted for the production augmenting capital stock, thus allowing a larger proportion of saving to be allocated to investments which increase the natural environment's assimilative capability. Likewise, logarithmic technical change can be assumed for either environmental investments or investments in production of goods and services. Such complications affect the interpretation of feasible positive and negative rates of growth, but add only a marginal increment to the viewpoints expressed here.

dissolved oxygen concentrations. Thus, the model is nonspecific regarding processes to augment the environment's assimilative capability.

The model suffers, however, in requiring extreme aggregation assumptions on different types of waste flow and in specifying a linear coefficient connecting assimilative investment to reductions in waste density. The latter is very likely to be not only non-linear, but non-additive in terms of different types of wastes. Also, no consideration was given to the possibility of recycling investments which not only reduce current waste flow but may also contribute to production as well.[1] However, the model has the advantage of being highly illustrative of the policy arguments that I now wish to consider. Many additional minor conclusions can be squeezed out, but this would induce additional pollution via the paper and pulp industry so I shall leave them for the reader.

What we have enumerated from this model is the fact that unregulated behavioral conditions such as the marginal propensity to save influence in the long-run whether a positive rate of growth in output is warranted when a constraint is imposed on the utilization of environmental assimilative capacity. Certainly in a private market economy, the common property character of almost all environmental resources and failure of legal systems to adequately define common property rights means that there is little or no individual incentive to invest in augmenting the environment's waste assimilative capability. In consequence, such economies will generally be plagued with overutilization of the natural environment, unless these economies are blessed with exceedingly large natural environments relative to the magnitude of wastes generated from production-consumption. But what is most important in terms of mixed public-private economies is the realization of the degree of interdependence between decisions on economic growth and the environment. Most certainly production-consumption and waste generation need not be strictly joint products provided the natural environment is treated by public authorities as an asset which needs to be considered in *all* other decisions with respect to the economy's development and stability. There are no neutral fiscal or monetary policies when environmental assets are considered. Either growth in consumption as well as rates of investment which augment the capacity for increasing physical product are encouraged or discouraged. In either case, there is a direct effect on the utilization of the natural environment's assimilative capability.

What the simple model emphasizes is the types of policy instruments the government has at its disposal to influence the natural environment beyond effluent charges, subsidies, or direct standards on waste emissions of industries and municipalities. First, not only are shifts in propensities to save likely to increase potential growth rates but also provide the necessary investment for

---

[1] One should note the possibility that certain types of recycling investment may cause very high waste loads per dollar of production, thus partially or completely offsetting recycling gains.

increasing the environment's capacity to assimilate wastes. Secondly, higher saving increases the time span between production and waste generation at least for some types of capital equipment investment. However, if assimilative investments are not undertaken, higher savings rates only aggravate the environmental problem in the long run. For a warranted positive rate of growth in output to be commensurate with no or little change in waste densities, a reasonably high efficiency must be obtained in assimilative investments.

A nationally consistent environmental policy means the integration and coordination of economic growth and stability with specific decisions on pollution. Not only are relative prices not reflecting discrepancies between marginal social and private costs due to environmental pollution, but governmental policies directed toward "pure growthmanship" are also inducing externalities by accelerating the generation and emission of wastes.

A fiscal policy that emphasizes maintenance of high levels of aggregate demand and employment quite innocently also accelerates rates of waste generation and accumulation if it is not coupled with an environmental investment program. Monetary policies aimed at easing credit have the simultaneous impact of increasing immediate waste flows via a positive shift in consumption expenditure but a negative impact through encouragement of investment and thus retardation in the flows of goods immediately consumed. Selective instruments such as investment tax credits, depletion allowances, and capital gains provisions also have direct and indirect effects on the rate of consumption and the distribution of production between current consumption and savings. Therefore these instruments also influence the potential rate of environmental deterioration.[1]

This discussion is not meant to be an analysis or even a complete statement on the effects of economic policies on the generation of wastes. However, I believe this most limited exposition suggests that macroeconomic policy instruments and decisions on economic development tactics have a definite and often unrecognized impact on the natural environment through their effects on rates of resource extraction, growth in consumption per capita, and disruption of natural food and energy chains.

Policies directed toward accelerating depreciation and even policies aimed at improving balance of payments influence rates of waste generation. In recent decades, there is some evidence the United States and several other developed nations have become or are becoming net importers of natural resources and net exporters of consumer durables which is indicative, on

---

[1] It is interesting to speculate whether the opponents of the oil depletion allowance in the United States considered the question whether such an allowance actually increased the rate of waste flows, particularly automotive emissions of carbon monoxide and oxides of nitrogen? The proponents for the Aswan Dam project most certainly neglected to analyze the problems resulting from nutrient imbalances downriver, with the consequent reduction in both soil fertility and fish yield.

balance, they may very well have become net importer's of waste emissions.[1] Most certainly the current terms of trade *do not* reflect long-term social costs (or disutilities) associated with the wastes emitted.

IV

One of the most perplexing difficulties confronting a rational utilization of the world environment is the effect environmental policies will have on the growth of developing and developed countries. It is difficult to even conceptualize the waste loads that would be exerted on the world environment if all countries had standards of living approaching those of the mature nations. Here, only one dimension of this problem will be briefly examined: namely, the effect on international comparative advantage of selected developed countries if these countries undertake a control program directed toward reducing or neutralizing their pollution loads.

Specifically, the establishment of effluent charges or standards to reduce industrial and municipal waste emissions will raise domestic costs of production, provided in-plant processing changes, substitution between factor inputs, recycling of coolants, and other changes induce additional expenditure by polluters. In part, such standards or charges should have the impact of shifting investment from augmenting production to directly or indirectly augmenting the natural environment's assimilative capability.[2] The expected domestic effect however will be reduced profits to relatively high waste load emitting industries and increased domestic and export prices. The extent of the division between higher prices and lower profits depends on such factors as market and cost structures.

If a nation *unilaterally* imposes higher waste emission standards on its domestic industries, a rise in domestic and export prices will usually result in placing the nation at an international disadvantage. For relatively open economies exporting highly competitive commodities internationally, the impact of such standards could spell a significant reduction in domestic income. This income reduction, of course, would reduce further the amount of waste flow generated by decreasing the rate of consumption. It may also have the impact of reducing the availability of investment for augmenting assimilative capability via the results of the growth model discussed earlier. Thus, a set of effluent charges which reflect current domestic social damages and are optimal from a domestic standpoint may have international repercussions which reduce waste flows to less than optimal levels domestically.[3]

---

[1] See for example, J. Vanek, *The Natural Resource Content of United States Foreign Trade: 1870–1955*, Cambridge, Massachusetts: MIT Press (1963).

[2] The effects of such shifts can be readily seen from the model discussed in Section III.

[3] In effect, the old transfer problem of international trade on over and under compensation has reared its ugly head in terms of the newer problem of environmental control. See J. M. Keynes, "The German Transfer Problem", *Economic Journal*, Vol. XXXIX (March, 1929); and B. Ohlin, "The Reparation Problem: A Discussion", *Economic Journal*, Vol. XXIX (June, 1929); or the modern synthesis in J. Vanek, *International Trade: Theory and Economic Policy*, Homewood, Illinois: Richard Irwin (1962).

Table 1. *Estimated change in domestic income resulting from the imposition of environmental controls, five countries, 1968 (billions of dollars)*[b]

Country	% change in domestic and export prices[a]	Projected negative change in domestic income	Projected % negative change in domestic income
United States	1.0–3.6	1.0– 3.6	0.1– 0.4
Japan	1.0–4.0	7.0–28.1	4.9–20.0
West Germany	1.0–4.0	0.9– 3.7	0.7– 2.7
United Kingdom	1.0–4.0	6.5–25.8	6.3–25.0
France	1.0–3.7	9.6–35.3	7.5–28.0

[a] Estimated range of changes in domestic and export prices resulting from this imposition of effluent charges or standards.
[b] These estimates must be viewed as purely experimental and preliminary as their magnitude and sign are extremely sensitive to slight changes in empirical specifications.

Alternatively, augmenting waste assimilative capability through general taxes by subsidizing industrial and municipal pollution control is likely to shift relative prices less but place the impact directly on lowering domestic real incomes. However, it is difficult to assess whether effluent charges or subsidies will have the greater impact via international advantage.

In Table 1 are recorded empirical estimates of the impact a significant shift in domestic and export prices induced by unilaterally imposed effluent charges and/or standards would have on domestic incomes for five selected countries. Many simplifying assumptions were necessary to make these calculations and they will not be documented in detail here. Briefly, a three equation model was utilized connecting each country with the "rest of the world" where import demand and domestic absorption were determined by domestic output and domestic terms of trade.[1] The model utilized was one which expli-

---

[1] The three equation model utilized to calculate the changes in income was of the form:

$$Y_i - M_{ii}(Y_i, T) - M_{ij}\left(Y_j, \frac{1}{T}\right) = 0$$

$$Y_j - M_{ji}(Y_i, T) - M_{jj}\left(Y_j, \frac{1}{T}\right) = 0$$

$$M_{ij} - \frac{1}{T} M_{ji} = B_i$$

where Y_i, M_{ii}, M_{ij}, and B_i are respectively: domestic output, domestic absorption, exports to the rest of the world, which is denoted as j, and balance of payments on current account of country i. T denotes the gross barter terms of trade of country i. Differentiating these three equations with respect to a change in domestic prices; solving for the change in income of country i; and inserting empirical estimates of import elasticities, domestic cross price elasticities and marginal propensities to import and consume, yielded estimates in Table 1. Published estimates were used for all elasticities and propensities except the cross elasticities of demand which were derived from our own single equation regression estimates. Weighted elasticities and propensities were used for the "rest of the world"

citly omitted the governmental sector, and therefore monetary policy is implicitly assumed to be operating in such a way that interest rates are stabilized and exchange rates are pegged. In essence then, the model does not consider *any* type of governmental policy which could be directed toward alleviating or eliminating the effect of domestic effluent charges on international advantage. Consequently, the crude estimates in Table 1 should be viewed as the most extreme effect on domestic incomes resulting from the imposition of effluent charges or standards.

The shift in domestic and export prices from a given change in environmental control which attempts to equate marginal private and social costs is difficult if not impossible to measure at the level of aggregation applied here. For reference purposes, I have utilized the range of from one percent to a percentage subjectively determined which would yield an "appreciable" reduction in waste load for individual industries. These percentages were then weighted according to the export composition for each of the countries listed in Table 1. Therefore, these ranges must be considered as pure judgments from limited experience on United States industrial pollution loads. The estimated percentage changes appear, however, to be biased downward for any relatively large scale national effort to reduce pollution loads.

The estimates in Table 1, although qualified with a long list of assumptions, lead to several interesting conjectures on the establishment of national and/or international pollutant emission controls. First, the effect of establishing environmental controls is very unlikely to be internationally neutral ... some countries, due to the type of product exported and variations in demand elasticities and marginal propensities, will undoubtedly suffer a greater "burden". This burden in terms of reduced domestic incomes will probably result in too large an adjustment to waste induced damages domestically. Second, countries which are relatively "open", in that exports account for a sizable proportion of domestic output, will tend to be affected more by impositions of environmental controls than economies, such as the United States, which are relatively closed. Economies confronted with elastic (price) demands for export commodities with relatively high pollution loads per unit of product produced and consumed will be subject to falling export revenues and reductions in domestic demand. In consequence, such economies undoubtedly face a greater burden than economies with less elastic demands and greater environmental assimilative capabilities such that effluent charges have less impact on export prices.

It should be recognized that economies may (and probably will) place high tariffs or quotas on imports which induce a relatively high waste load

where the weights corresponded to the volume of imports to country *i* from other countries. Also, application of the three equation model above implies that *all* commodities are produced under conditions of constant marginal cost. For a complete discussion of the procedure applied see, R. C. d'Arge "Environmental Standards and Comparative Advantage", unpublished manuscript (November, 1970).

during or following their "consumption" even if defensive action via tariffs directed toward alleviating the effect of domestic effluent charges is not undertaken. Such a set of tariffs is not only necessary for protecting domestic environmental quality but also for free international trade to result in global type of efficiencies.[1]

The final consequences of nationally imposed controls on each country cannot be foretold since they depend not only on price elasticities, income propensities, and domestic substitutions in production, but also on publically induced tariff, quota, and indirect measures to offset environmental controls. What I have attempted to illustrate is not the particular policy apparatus which will affect comparative advantage but only that pollution control efforts can potentially have a marked effect on comparative international advantage of particular countries. If we accept these rather fragmentary bits of evidence, then the conclusion follows again that governmental decisions at the national and international level are influencing and therefore ought to be influenced by considerations of the interaction of man and the environment.

V

Several threads of thought have permeated each section of this essay and it seems appropriate to draw them together here. Production, consumption, and waste generation are joint products except for the activities of man toward recycling his own or the environment's wastes, his actions toward augmenting the natural environment's assimilative capability through investments, or his actions toward reducing his own demands for material goods. In the very long run, mankind may be faced with a finite amount of resources and an infinite planning horizon ... Nietzsche confronted this problem in the doctrine of Eternal Recurrence.[2] If a finite planning interval is assumed along with some unsevere assumptions regarding utility, disutility, and population change, minimum current consumption appears to be appropriate.

A declining rate of current consumption and output was also found to be consistent with the steady state assumption on environmental pollution, provided the saving and assimilative investment coefficients of an economy were not large enough to compensate for the economy's rate of waste generation. Economic growth and environmental quality are only compatible in the long run provided that as growth in output occurs, a significant proportion

---

[1] For a discussion of the assumptions necessary for world efficiency in terms of free trade, see: P. A. Samuelson, "The Gains from International Trade", *Canadian Journal of Economics and Political Science*, Vol. V (May, 1939); P. A. Samuelson, "The Gains from International Trade Once Again", *Economic Journal*, Vol. LXXII (December, 1962); or the excellent discussion in M. C. Kemp, *The Pure Theory of International Trade and Investment*, Englewood Cliffs, New Jersey: Prentice Hall (1969), pp. 253–285.

[2] See, A. C. Danto, *Nietzsche as Philosopher*, New York: Macmillan (1965) pp, 205–206.

of investment is directed toward recycling wastes or increasing the assimilative capability of the environment. These findings indicate to me the necessity of analyzing very carefully policies directed toward stabilizing and even reducing per capita incomes in the "mature" economies.

The effects of environmental controls on international advantage of particular economies appear to be quite pervasive, even if one heavily discounts for all of the qualifications necessary to justify the empirical methods employed in estimation. This is especially true for relatively trade dependent nations. Economic policies directed toward the traditional targets of growth in output, stability, full employment, and the balance of payments all have seemingly unrelated but potentially destructive interactions with the natural environment and therefore cannot be separated from *any* national (or international) environmental policy.

ENVIRONMENTAL CONTROL AND ECONOMIC SYSTEMS

*Erik Dahmén**

The Stockholm School of Economics, Stockholm, Sweden

Aims and Limitations

This paper starts out by drawing attention to some basic facts regarding environmental deterioration in countries with different economic systems. From this it proceeds to deal with the appropriateness of approaches to solving environmental problems under the different systems.

There are three important limitations: First, only those forms of environmental damage which influence human well-being will be discussed. The economics of natural resources will only be included indirectly. Second, only industrialized countries will be dealt with. Third, two rather refined economic systems will be in focus: *Private capitalism* characterized by a market economy, i.e. freedom for separate entities to buy and sell competitively in a climate ruled by free price formation and profit maximization, and *socialism* without this freedom but characterized instead by centralized decisions on resource allocation and prices, including the setting of production goals for separate production units and, to a varying degree, incentives to attain various specific goals, such as minimization of production costs.

Historic Facts

The first question to be answered is: Has the pace of environmental deterioration been different under various economic systems? The answer appears to be no. Differences in the degree of environmental disruption exist between countries of different levels of development with different population densities, different degrees of urbanization, different geographical conditions and different climates. But there is no evidence whatsoever that the economic systems have been of any importance in this connection. It should be added—without discussing the point in detail—that neither the political organization, whether democratic or totalitarian, nor the political color of the government, appear to have been significant. Moreover, state-owned companies

---

* This paper is a revised edition of a Background Paper presented at the International Symposium on Environmental Disruption in the Modern World, organized in Tokyo by International Social Science Council, 8–14 March 1970.

have been no different from privately owned companies in the extent to which they have damaged the environment.

The fact that so far there has not been any connection between various economic systems and the rate of environmental deterioration, does not exclude the possibility that different systems might have different means of pursuing a successful environmental policy. This is because development, essentially similar, up to now, may have been due to inadequate knowledge. This in fact seems to be the main explanation for the absence of any connection between economic system and environmental deterioration. An additional explanation could be that the same comparatively low value has been placed on environmental conditions in both economic systems. Consumption has perhaps been given universal priority at the expense of the environment, particularly in countries which lag behind with respect to consumption, i.e. the socialist ones that started out on a low consumption level. But in most instances the damage has no doubt been tacitly accepted. Discussions leading to explicit decisions on the order of priorities have been the exception.

Causes of Environmental Deterioration

An analysis of the causes of environmental deterioration could serve as a useful basis for examining environmental control problems in various economic systems.

Attention could be drawn to technological and industrial progress, population development, urbanization, geographical conditions and climate. But technology, industrialization and urbanization themselves are the results of more basic factors. Geographical conditions and climate can, at most, be regarded as contributory causes, but never sufficient in themselves to cause environmental degradation. However, if we take a step backward in the causal chain, we find that certain characteristic conditions have played a determining role everywhere.

One condition is organizational in nature, i.e. a highly developed disintegration of economic activities into discrete stages that was unknown in old agrarian societies. Different stages of decision-making have arisen to take charge of different parts of the production process and very often decisions within separate producing units have today far-reaching consequences for the general public which can have very little influence on the decisions. This organizational structure has been created by the process of industrialization in both main types of economic system. In the capitalistic system, no production goals for separate producers are set by a central authority. On the whole, no central directives are given with respect to the organization of production, the choice of a production process, financing measures and so on—or to questions of pollution and waste treatment. In socialistic systems, central authorities set production goals and decide on questions of resource allocation and financing, but producers have approximately the same freedom as in a capitalistic system in the area of technological and other types of measures affecting the environment.

This disintegration has been accompanied by the creation of a growth-promoting network of continuous contacts among producers and between producers and consumers. These contacts have been established either through the medium of a price mechanism acting on a free market or have been organized by special institutions allocating real and financial resources as well as final products among the separate entities. The first method has been developing gradually in private capitalistic countries, whereas the second was established by means of political decisions in socialistic countries.

Besides these two types of contact networks disintegration has had numerous other effects some of which are of great importance not only from a theoretical point of view but also with respect to environmental developments.

One important effect is referred to in the literature dealing with market economies as "externalities" defined as effects of production and consumption on other producers and consumers without corresponding payments. These externalities can be either positive or negative. The negative ones include environmental damages which imply a use of scarce resources and, consequently, should be regarded as socio-economic costs. Such damages represent externalities partly because environmental utilities often are so-called "public goods" i.e. goods which if available to one are equally available to all others and therefore cannot be sold and bought on a market. Thus in market economies no "bills" are sent to producers and consumers through the market mechanism for this kind of socio-economic costs. Moreover, particularly under strong competitive conditions, considerations of the profit-and-loss-account sharply limit the producers' possibilities of calculating, voluntarily, the negative externalities as costs. So far governments have corrected this "market failure" only in very limited ways and to a slight extent.

In socialistic systems characterized by the same kind of disintegrated activities, environmental damages, being "external" from the viewpoint of the separate entities, have generally been neglected in the same way as in the private capitalistic market economies. In other words: Externalities which by definition cannot possibly be taken care of by the market mechanism have not been taken care of by any other mechanism in the non-market economies. In the latter case there has been a "planning failure".

Under both systems, a number of remarkable consequences of this failure of the market and/or the authorities to react have appeared.

One consequence has been that producers have regarded production methods and products which damage the environment as the least expensive alternatives. If there had been no possibility of omitting the socio-economic costs represented by environmental damages from the cost calculations, these alternatives would instead have often appeared to be the most expensive.

Thus cost calculations that are too narrow have eliminated economic incentives to choose non-damaging production methods and products as long as the damaging ones are cheaper.

As a result of the choices made by producers, prices of goods whose production, distribution or use involved environmental degradation have been too low with respect to socio-economic costs. Consequently, consumption and production of goods such as these have been stimulated in a way which damages the environment, as compared with the production and consumption of goods whose production, distribution or use has not damaged the environment. Indirect effects have thus been added to the direct ones.

These two immediate consequences have in turn had another effect, the importance of which has been considerably underestimated not only by non-economists but also by many economists who are primarily occupied with theories applicable to stationary economies or who regard technological progress as an exogenous variable in the process of industrial development. Since technological research is often influenced by business considerations, this type of research has shown comparatively little interest in finding and developing new techniques and new products with less damaging effects on the environment. This lack of incentives to *innovate* methods and products beneficial to the environment is one of the most important, but very much neglected, causes behind long-run environmental deterioration. Furthermore, there are reasons to believe that a misallocation of scarce resources for research has implied lost opportunities of limiting environmental degradation at a considerably lower cost than has so far proved possible.

These factors, all of which are not basically related to the ownership of productive means nor to the market or non-market organization but to a great extent to the disintegration of activities and thus to the existence of externalities, have formed the basis for the environmental deterioration that has occurred everywhere. In both economic systems certain additional technological elements have entered the picture i.e. those leading to numerous *large* production units. Therefore, environmental damage has quantitatively become very significant and has been spread over vast geographical areas. As a result, legal redress for damages has proved increasingly difficult to obtain. Environmental disruption has appeared instead as a general welfare problem for society as a whole.

Goals and Means of an Environmental Policy: A Theoretical Approach

Starting from this sketchy analysis of the factors behind environmental disruption let us first deal with the problems related to the way environmental policy *goals* may be set. Considering the fact that something (i.a. consumption) has to be sacrificed in order to achieve an environmental improvement, a basic question becomes *who* should decide on the goals, i.e. *how much* is it desirable to sacrifice in favor of a better environment? The question is whether different economic systems have essentially different characteristics, i.e. whether they face different difficulties with respect to the setting of the goals.

I have difficulty in seeing any difference worth mentioning between the two economic systems in this respect. A possible desire to accept consumer preferences would make it necessary in both cases to rely on some sort of opinion poll. If there is no free market, this is self-evident. But a free market which in a sense is usually a mechanism functioning continuously as an "opinion poll" does not solve the problem either. Many vital environmental values cannot be registered by the price-system revealing consumer preferences, since they are "public goods". However, opinion polls on environmental problems would be difficult to carry out, regardless of the economic system. There are obvious reasons for this.

Whenever the question of environmental improvement comes up a position must be taken on the goals of income distribution or, rather, on distribution of welfare. The benefits of a measure that improves the environment will by no means always be enjoyed in equal measure by those who have to pay for it. This question of welfare distribution also has an intertemporal aspect in that several generations are affected. This is especially important in the area of the environment and it renders the idea of building on "general opinion" rather suspect. This is particularly because many forms of environmental damage may be irreversible.

In addition policies to improve the environment may sometimes give rise to problems with respect to the balance of payments. Although problems such as these could always be solved by using a package of policy means it is easy to imagine the difficulties in having people take a stand with regard to matters that are often highly complicated and "technical".

Even quite apart from this type of problem, it should be pointed out that public opinion could seldom be based on sufficient information of a generally understandable nature. Questions of fact are often subject even to scientific controversy—for example in reference to the forms and magnitudes of risks in environmental deterioration.

The general conclusion seems unescapable that the goals of environmental control must on the whole be based on a factual basis, preferably by means of some sort of cost-benefit-analysis, available only to central decision-making bodies. Very often it will be a question of making investment calculations under great uncertainties. In this respect there are no significant differences between capitalistic market and socialistic non-market systems.

Let us then turn to *means* of improving the environment in the different economic systems. The first observation is as follows: In view of the fact that environmental deterioration is basically caused by the externalities characteristic of all systems of disintegrated activities one way of attaining environmental control in both systems stands out fairly clearly in relation to the "resource allocation approach" to the problem of environmental degradation. It would involve having decision-makers within the separate entities weigh damage to the environment as a cost just like any other cost of doing business.

One question concerns whether this problem can be solved through centralization of the *decision-making process* as regards choices of production methods, product design, pollution, waste treatment, etc. Is it possible, assuming retention of the disintegration of activities, to establish one or relatively few decision-making instances where some of the units' externalities can be "internalized", i.e. made to enter into their cost calculations?

In a private capitalistic system, such a strong centralization of decisions is not conceivable. Producers cannot be guided in every separate instance by detailed central directives on all technical points involved in production methods and products. Both the legal and the administrative difficulties would be too great.

Establishing one or more large centralized decision-making bodies may appear more practicable in a socialistic system, inasmuch as ownership itself is already centralized. But this kind of centralized authority is not practicable here either. The legal aspects offer no problems, but the administrative side does. Even here producers cannot be given instructions on every aspect of the problems on a case-to-case basis.

This leaves us with two more conceivable ways of attacking the environmental policy problem. There are two main approaches worth discussing.

One group of means (usually preferred by legal experts, administrators and other laymen in the field of economics) consists of prohibitions or agreements which are referred to here simply as *regulations*. These regulations mean that the socio-economic costs represented by the damages to the environment are not permitted to exceed certain limits, which are determined from case to case. The rules can be unspecific so that the producer in question can, himself, attempt to discover how to limit the damage. But the rules can also give instructions on how the delimitations should occur.

One result of this approach is that producers' costs are raised. They may be forced to invest, for example in waste treatment plants. Or they may shift to more expensive production or distribution methods. In both cases the environment is improved not only directly but also indirectly. The price increases which producers can be presumed to try are likely to reduce demand and therefore lead to a reduction of production that still damages the environment, though to a smaller degree than earlier.

Such regulations, based on intermittent administrative initiatives from instances outside the producer circles, can be used in both economic systems. The question is whether the method might be expected to be equally effective in both systems.

In a market economy a desire arises on the part of producers to bring down the costs of a waste treatment plant or a new production process in order to reduce the strain on the profit-and-loss account. In non-market economies regulations can first of all be presumed to make it more difficult to reach a prescribed production goal. Therefore a similar incentive would also appear here, if the production goals set by the central planners are maintained and no sufficiently increased supply of resources is made available, such as additional financing. The question is whether the incentives in such a case can be expected to function with as much force as when profits are threatened in a private capitalistic market system. The answer depends on the effectiveness of the incentives for managements to minimize production costs. Therefore one crucial question is how i.a. bonuses used in such a system really function in this respect. In my opinion there is not much evidence of effective functioning so far. But this does not exclude the possibilities of introducing appropriate bonuses for attaining specific goals with respect to environment quality.

One characteristic of all regulations in both economic systems is that they are rather inflexible and not very well suited for reducing the damages to an optimum level from a general resource allocation point of view. Furthermore there is no incentive among producers to find *new* production methods and products more beneficial to the environment but not prescribed by the regulation. This incentive to innovate would be limited to firms offering various abatement and purification equipment and alternate processing techniques and possibly cooperating with those who have directly or indirectly the regulative power. The incentive for the other producers can be assumed to be directed at just minimizing the extra costs brought about by the regulation. Even this limited incentive would disappear if the authorities choose to subsidize away the marginal cost increases resulting from the regulative prescription. Furthermore, in such a case there would be no corrections of the price relations between various products and thus no indirect beneficial effects on the environment. This would be irrational from the point of view of environmental policy.

Owing to the limitations and weaknesses of the regulation method there are reasons to look for another method which, at least with respect to the points touched on so far, would be superior to the regulation method. And, if so, how useful might it be in different economic systems?

The approach which is most immediately apparent to an economist is that which involves *charging a fee* for environmental damages. This method would not set any strict limitations for the damages inflicted on the environment. Instead, a bill would be presented. Its amount would be reduced or increased proportional to the reduction or increase in environmental damage. This can be assumed to help reduce the damages to an optimum level from a general resource allocation point of view in a more effective way than the regulation method. The fact that a stronger incentive to bring down the damage would be given is of particular importance. Even if the initial effect in the form of, for example, investment in waste treatment equipment or other technical solutions might not be the same in the fee method as in the regulation method (i.e. in the case of rather small fees) there is a great probability that the effect in terms of environmental improvement over a *period of time* would be greater. Above all it could be hoped that *technological progress* in the field of environmental control would receive a powerful and continuous stimulus. In theories of economic development, insufficient attention is usually paid to the fact that impulses go out from company managements and from potential founders of enterprises to technicians, product designers and indirectly to researchers. It would be of great benefit, particularly in a discussion of environmental policy means, if we could include this fact in the picture. I would launch the hypothesis that through innovation in various abatement and treatment techniques, in processing and in the form of new products, consumers' sacrifices in terms of consumption could be reduced considerably, possibly in the short run but certainly over the *long term*, by the use of fees as an important complement to regulations. It seems quite possible that the marginal returns on research and development investments on the environmental front could be large. It is even likely that basic research, which is not stimulated by industrial activity but develops by reason of "pure curiosity", has long possessed a store of potentially valuable technology to offer in the environmental area but that it has not yet been of special interest for producers to find and develop these possibilities.

Now the question is how strong the incentives provided by the fee method would be in a market economy compared to a non-market economy.

In a private capitalistic market economy the incentives no doubt would prove strong because of the strain on profit and loss-accounts. In the other economic systems, the fees are likely to give comparatively weak incentives and to be less effective than regulations. This is because the fees cannot usually make attainment of a prescribed production goal more difficult in the same way as regulations. The effects of a fee could be absorbed more easily, for example by cutting outlays in some other areas in a way that would not immediately disturb the factory's production to the same degree. This conclusion should be modified only insofar as there is an effective "profit" incentive even in the non-market system, i.e. to the extent that a factory's possible cost reductions are generally fully rewarded. So far, this does not seem to have occurred. Here again, however, future improvements i.a. in the bonus system might change the picture.

Before arriving at some summary conclusions it should be kept in mind that comparisons here have been made between rather refined types of economic systems and equally refined alternatives for environmental control. This has been done exclusively in the interest of simplification. Such a method may be acceptable as a basis for an analytically more ambitious and more detailed discussion of various "mixed" economic systems and various combinations of policy means.

Summary Conclusions

1. Since severe environmental deterioration has been going on for a long time regardless of economic system, measures are now discussed both in capitalistic market economies and in socialistic non-market economies to bring about a shift in direction. If we draw up a static picture of today's situation and that of the future, there is reason for pessimism, especially since all countries have a costly bill to settle for many years of sins. But if we take a more dynamic view of the matter and if we succeed in working out rational methods for environment policies, then there are grounds for optimism. There is no natural law saying that present environmental disruption is unavoidable and the costs for correcting damages might not, over the long term, prove to be as enormous as they may seem today.

2. Environmental policy measures are surely available in both economic systems which I have here dealt with. The central problem does not lie in the conditions of ownership nor in the market or non-market organization but in the disintegration of activities which has been an element in industrialization and therefore in economic growth everywhere. This disintegration has resulted in calculations of socio-economic costs that are too narrow and therefore in a misallocation of resources and environmental deterioration. The main task of environmental control should be to enforce more accurate cost calculations.

3. More accurate cost calculations would make producers inclined to choose technical alternatives more favorable to preserving the environment. Some such technical possibilities already exist although it has not been attractive to use them. More accurate cost calculations would also influence future technology in the right direction. Therefore there is in both economic systems a hitherto neglected opportunity to stimulate technology in a way more beneficial to the human environment than we have experienced so far.

4. In capitalistic market economies the most effective way of enforcing changed cost calculations and, in particular, of promoting new technologies would very likely be to *charge* producers for the damage done to the environment. This is because charges such as these would create more or less strong business incentives to act in compliance with the environmental policy goals set up by authorities. In a socialistic non-market system the impact of charges for environmental damage would be dependent on the extent to which separate units are under pressure not only to reach prescribed production goals but also to minimize production costs and maximize "profits". Insofar as such pressures are comparatively weak, *regulations* of various kinds instead of charges would seem more appropriate.

THE USE OF STANDARDS AND PRICES FOR PROTECTION OF THE ENVIRONMENT

*William J. Baumol and Wallace E. Oates**

Princeton University, Princeton, N.J., USA

Summary

In the Pigouvian tradition, economists have frequently proposed the adoption of a system of unit taxes (or subsidies) to control externalities, where the tax on a particular activity is equal to the marginal social damage it generates. In practice, however, such an approach has rarely proved feasible because of our inability to measure marginal social damage.

This paper proposes that we establish a set of admittedly somewhat arbitrary standards of environmental quality (e.g., the dissolved oxygen content of a waterway will be above x per cent at least 99 per cent of the time) and then impose a set of charges on waste emissions sufficient to attain these standards. While such *resource-use prices* clearly will not in general produce a Pareto-efficient allocation of resources, it is shown that they nevertheless do possess some important optimality properties and other practical advantages. In particular, it is proved that, for any given vector of final outputs such prices can achieve a specified reduction in pollution levels at minimum cost to the economy, even in the presence of firms with objectives other than that of simple profit maximization.

In the technicalities of the theoretical discussion of the tax-subsidy approach to the regulation of externalities, one of the issues most critical for its application tends to get the short end of the discussion. Virtually every author points out that we do not know how to calculate the ideal Pigouvian tax or subsidy levels in practice, but because the point is rather obvious rarely is much made of it.

This paper reviews the nature of the difficulties and then proposes a substitute approach to the externalities problem. This alternative, which we shall call the environmental pricing and standards procedure, represents what we consider to be as close an approximation as one can generally achieve in practice to the spirit of the Pigouvian tradition. Moreover, while this method does not aspire to anything like an optimal allocation of resources, it will be shown to possess some important optimality properties.

---

* The authors are members of the faculty at Princeton University. They are grateful to the Ford Foundation whose support greatly facilitated the completion of this paper.

1. Difficulties in Determining the Optimal Structure of Taxes and Subsidies

The proper level of the Pigouvian tax (subsidy) upon the activities of the gene-rator of an externality is equal to the marginal net damage (benefit) produced by that activity.[1] The difficulty is that it is usually not easy to obtain a reasonable estimate of the money value of this marginal damage. Kneese & Bower report some extremely promising work constituting a first step toward the estimation of the damage caused by pollution of waterways including even some quantitative evaluation of the loss in recreational benefits. However, it is hard to be sanguine about the availability in the foreseeable future of a comprehensive body of statistics reporting the marginal net damage of the various externality-generating activities in the economy. The number of activities involved and the number of persons affected by them are so great that on this score alone the task assumes Herculean proportions. Add to this the intangible nature of many of the most important consequences—the damage to health, the aesthetic costs—and the difficulty of determining a money equivalent for marginal net damage becomes even more apparent.

This, however, is not the end of the story. The optimal tax level on an ex-ternality generating activity is not equal to the marginal net damage it gener-ates *initially*, but rather to the damage it would cause if the level of the activity had been adjusted to its *optimal* level. To make the point more specifically, suppose that each additional unit of output of a factory now causes 50 cents worth of damage, but that after the installation of the appropriate smoke-control devices and other optimal adjustments, the marginal social damage would be reduced to 20 cents. Then a little thought will confirm what the appropriate mathematics show: the correct value of the Pigouvian tax is 20 cents per unit of output, that is, the marginal cost of the smoke damage *corresponding to an optimal situation*. A tax of 50 cents per unit of output corresponding to the current smoke damage cost would lead to an excessive reduction in the smoke-producing activity, a reduction beyond the range over which the marginal benefit of decreasing smoke emission exceeds its marginal cost.

The relevance of this point for our present discussion is that it compounds enormously the difficulty of determining the optimal tax and benefit levels. If there is little hope of estimating the damage that is currently generated, how much less likely it is that we can evaluate the damage that would occur in an optimal world which we have never experienced or even described in quantitative terms.

There is an alternative possibility. Instead of trying to go directly to the optimal tax policy, one could instead, as a first approximation, base a set of

---

[1] We will use the term marginal *net* damage to mean the difference between marginal social and private damage (or cost).

taxes and subsidies on the current net damage (benefit) levels. Then as outputs and damage levels were modified in response to the present level of taxes, the taxes themselves would in turn be readjusted to correspond to the new damage levels. It can be hoped that this will constitute a convergent, iterative process with tax levels affecting outputs and damages, these in turn leading to modifications in taxes, and so on. It is not clear, however, even in theory, whether this sequence will in fact converge toward the optimal taxes and resource allocation patterns. An extension of the argument underlying some of Coase's illustrations can be used to show that convergence cannot always be expected. But even if the iterative process were stable and were in principle capable of yielding an optimal result, its practicality is clearly limited. The notion that tax and subsidy rates can be readjusted quickly and easily on the basis of a fairly esoteric marginal net damage calculation does not seem very plausible. The difficulty of these calculations has already been suggested, and it is not easy to look forward with equanimity to their periodic revision, as an iterative process would require.

In sum, the basic trouble with the Pigouvian cure for the externalities problem does not lie primarily in the technicalities that have been raised against it in the theoretical literature but in the fact that we do not know how to determine the dosages that it calls for. Though there may be some special cases in which one will be able to form reasonable estimates of the social damages, in general we simply do not know how to set the required levels of taxes and subsidies.

2. The Environmental Pricing and Standards Approach

The economist's predilection for the use of the price mechanism makes him reluctant to give up the Pigouvian solution without a struggle. The inefficiencies of a system of direct controls, including the high real enforcement costs that generally accompany it, have been discussed often enough; they require no repetition here.

There is a fairly obvious way, however, in which one can avoid recourse to direct controls and retain the use of the price system as a means to control externalities. Simply speaking, it involves the selection of a set of somewhat arbitrary standards for an acceptable environment. On the basis of evidence concerning the effects of unclean air on health or of polluted water on fish life, one may, for example, decide that the sulfur-dioxide content of the atmosphere in the city should not exceed x percent, or that the oxygen demand of the foreign matter contained in a waterway should not exceed level y, or that the decibel (noise) level in residential neighborhoods should not exceed z at least 99 % of the time. These acceptability standards, x, y and z, then amount to a set of constraints that society places on its activities. They represent the decision-maker's subjective evaluation of the minimum standards that must be met in order to achieve what may be described in persuasive terms as "a reasonable quality of life". The defects of the concept will immediately be clear to the reader, and, since we do not want to minimize them, we shall examine this problem explicitly in a later section of the paper.

For the moment, however, we want to emphasize the role of the price system in the implementation of these standards. The point here is simply that the public authority can levy a uniform set of taxes which would in effect constitute a set of prices for the private use of social resources such as air and water. The taxes (or prices) would be selected so as to achieve specific acceptability standards rather than attempting to base them on the unknown value of marginal net damages. Thus, one might tax all installations emitting wastes into a river at a rate of $t(b)$ cents per gallon, where the tax rate, t, paid by a particular polluter, would, for example, depend on b, the BOD value of the effluent, according to some fixed schedule.[1] Each polluter would then be given a financial incentive to reduce the amount of effluent he discharges and to improve the quality of the discharge (i.e., reduce its BOD value). By setting the tax rates sufficiently high, the community would presumably be able to achieve whatever level of purification of the river it desired. It might even be able to eliminate at least some types of industrial pollution altogether.[2]

Here, if necessary, the information needed for iterative adjustments in tax rates would be easy to obtain: if the initial taxes did not reduce the pollution of the river sufficiently to satisfy the preset acceptability standards, one would simply raise the tax rates. Experience would soon permit the authorities to estimate the tax levels appropriate for the achievement of a target reduction in pollution.

One might even be able to extend such adjustments beyond the setting of the tax rates to the determination of the acceptability standards themselves. If, for example, attainment of the initial targets were to prove unexpectedly inexpensive, the community might well wish to consider making the standards stricter.[3] Of course, such an iterative process is not costless. It means that at least some of the polluting firms and municipalities will have to adapt their

---

[1] BOD, biochemical oxygen demand, is a measure of the organic waste load of an emission. It measures the amount of oxygen used during decomposition of the waste materials. BOD is used widely as an index of the quality of effluents. However, it is only an approximation at best. Discharges whose BOD value is low may nevertheless be considered serious pollutants because they contain inorganic chemical poisons whose oxygen requirement is nil because the poisons do not decompose. See Kneese and Bower on this matter.

[2] Here it is appropriate to recall the words of Chief Justice Marshall, when he wrote that "The power to tax involves the power to destroy" (McCulloch vs. Maryland, 1819). In terms of reversing the process of environmental decay, we can see, however, that the power to tax can also be the power to restore.

[3] In this way the pricing and standards approach might be adapted to approximate the Pigouvian ideal. If the standards were revised upward whenever there was reason to believe that the marginal benefits exceeded the marginal costs, and if these judgments were reasonably accurate, the two would arrive at the same end product, at least if the optimal solution were unique.

operations as tax rates are readjusted. At the very least they should be warned in advance of the likelihood of such changes so that they can build flexibility into their plant design, something which is not costless (See Hart). But, at any rate, it is clear that, through the adjustment of tax rates, the public authority can realize whatever standards of environmental quality it has selected.

3. Optimality Properties of the Pricing and Standards Technique

While the pricing and standards procedure will not, in general, lead to Pareto-efficient levels of the relevant activities, it is nevertheless true that the use of unit taxes (or subsidies) to achieve the specified quality standards does possess one important optimality property: it is the least-cost method to realize these targets.[1] A simple example may serve to clarify this point. Suppose that it is decided in some metropolitan area that the sulfur-dioxide content of the atmosphere should be reduced by 50 %. An obvious approach to this matter, and the one that often recommends itself to the regulator, is to require each smoke-producer in the area to reduce his emissions of sulfur dioxide by the same 50 %. However, a moment's thought suggests that this may constitute a very expensive way to achieve the desired result. If, at existing levels of output, the marginal cost of reducing sulfur-dioxide emissions for Factory A is only one-tenth of the marginal cost for Factory B, we would expect that it would be much cheaper for the economy as a whole to assign A a much greater decrease in smoke emissions than B. Just how the least-cost set of relative quotas could be arrived at in practice by the regulator is not clear, since this obviously would require calculations involving simultaneous relationships and extensive information on each polluter's marginal-cost function.

It is easy to see, however, that the unit-tax approach can *automatically* produce the least-cost assignment of smoke-reduction quotas without the need for any complicated calculations by the enforcement authority. In terms of our preceding example, suppose that the public authority placed a unit tax on smoke emissions and raised the level of the tax until sulfur-dioxide emissions were in fact reduced by 50 %. In response to a tax on its smoke emissions, a cost-minimizing firm will cut back on such emissions until the marginal cost of further reductions in smoke output is equal to the tax. But, since all economic units in the area are subject to the same tax, it follows that the marginal cost of reducing smoke output will be equalized across all activities. This implies that it is impossible to reduce the aggregate cost of the specified decrease in smoke emissions by re-arranging smoke-reduction quotas: any alteration in this pattern of smoke emissions would involve an increase in

---

[1] This proposition is not new. While we have been unable to find an explicit statement of this result anywhere in the literature, it or a very similar proposition has been suggested in a number of places. See, for example, Kneese & Bower, Chapter 6, and Ruff, p. 79.

smoke output by one firm the value of which to the firm would be less than the cost of the corresponding reduction in smoke emissions by some other firm. For the interested reader, a formal proof of this least-cost property of unit taxes for the realization of a specified target level of environmental quality is provided in an appendix to this paper. We might point out that the validity of this least-cost theorem does not require the assumption that firms are profit-maximizers. All that is necessary is that they minimize costs for whatever output levels they should select, as would be done, for example, by a firm that seeks to maximize its growth or its sales.

The cost saving that can be achieved through the use of taxes and subsidies in the attainment of acceptability standards may by no means be negligible. In one case for which comparable cost figures have been calculated, Kneese & Bower (p. 162) report that, with a system of uniform unit taxes, the cost of achieving a specified level of water quality would have been only about half as high as that resulting from a system of direct controls. If these figures are at all representative, then the potential waste of resources in the choice between tax measures and direct controls may obviously be of a large order. Unit taxes thus appear to represent a very attractive method for the realization of specified standards of environmental quality. Not only do they require relatively little in the way of detailed information on the cost structures of different industries, but they lead automatically to the least-cost pattern of modification of externality-generating activities.

4. Where the Pricing and Standards Approach is Appropriate

As we have emphasized, the most disturbing aspect of the pricing and standards procedure is the somewhat arbitrary character of the criteria selected. There does presumably exist some optimal level of pollution (i.e., quality of the air or a waterway), but in the absence of a pricing mechanism to indicate the value of the damages generated by polluting activities, one knows no way to determine accurately the set of taxes necessary to induce the optimal activity levels.

While this difficulty certainly should not be minimized, it is important at the outset to recognize that the problem is by no means unique to the selection of acceptability standards. In fact, as is well known, it is a difficulty common to the provision of nearly all public goods. In general, the market will not generate appropriate levels of outputs where market prices fail to reflect the social damages (or benefits) associated with particular activities. As a result, in the absence of the proper set of signals from the market, it is typically necessary to utilize a political process (i.e., a method of collective choice) to determine the level of the activity.[1] From this perspective, the selec-

---

[1] As Coase and others have argued, voluntary bargains struck among the interested parties may in some instances yield an efficient set of activity levels in the presence of externalities. However, such coordinated, voluntary action is typically possible only in small groups. One can hardly imagine, for example, a voluntary bargaining process involving all the persons in a metropolitan area and resulting in a set of payments that would generate efficient levels of activities affecting the smog content of the atmosphere.

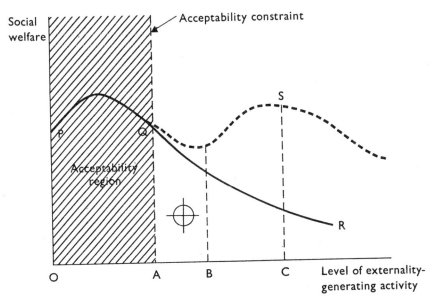

Fig. 1

tion of environmental standards can be viewed as a particular device utilized in a process of collective decision-making to determine the appropriate level of an activity involving external effects.

Since methods of collective choice, such as simple-majority rule or decisions by an elected representative, can at best be expected to provide only very rough approximations to optimal results, the general problem becomes one of deciding whether or not the malfunction of the market in a certain case is sufficiently serious to warrant public intervention. In particular, it would seem to us that such a blunt instrument as acceptability standards should be used only sparingly, because the very ignorance that serves as the rationale for the adoption of such standards implies that we can hardly be sure of their consequences.

In general, it would seem that intervention in the form of acceptability standards can be utilized with any degree of confidence only where there is clear reason to believe that the existing situation imposes a high level of social costs *and* that these costs can be significantly reduced by feasible decreases in the levels of certain externality-generating activities. If, for example, we were to examine the functional relationship between the level of social welfare and the levels of particular activities which impose marginal net damages, the argument would be that the use of acceptability standards is justified only in those cases where the curve, over the bulk of the relevant range, is both decreasing and steep. Such a case is illustrated in Fig. 1 by the curve *PQR*. In a case of this kind, although we obviously will not have an accurate knowledge of the relevant position of the curve, we can at least have some assurance that the selection of an acceptability standard and the imposition of a unit tax sufficient to realize that standard will lead to an increase in social welfare. For example, in terms of the curve PQR in Fig. 1, the levying of a tax sufficient to reduce smoke outputs from level OC to OA to ensure that the quality of the air meets the specified environmental standards would obviously increase social welfare.[1]

On the other hand, if the relationship between social welfare and the level of the externality-generating activity is not monotonically decreasing, the changes resulting from the imposition of an acceptability standard (e.g., a move from S to Q in Fig. 1) clearly may lead to a reduction in welfare. Moreover, even if the function were monotonic but fairly flat, the benefits achieved might not be worth the cost of additional intervention machinery that new legislation requires, and it would almost certainly not be worth the risk of acting with highly imperfect, inconclusive information.

In some cases, notably in the field of public utility regulation, some economists have criticized the employment of acceptability standards on both these grounds; they have asserted that the social costs of monopolistic misallocation of resources are probably not very high (i.e., the relevant portion of the social-welfare curve in Fig. 1 is not steep) and that the regulation can itself introduce inefficiencies in the operations of the regulated industries.

Advocacy of environmental pricing and standards procedures for the control of externalities must therefore rest on the belief that in this area we do have a clear notion of the general shape of the social welfare curve. This will presumably hold true where the evidence indicates, first that a particular externality really does have a substantial and unambiguous effect on the quality of life, if, for example, it makes existence very unpleasant for everyone or constitutes a serious hazard to health; and second that reductions in the levels of these activities do not themselves entail huge resource costs. On the first point, there

---

[1] The relationship depicted in Fig. 1 is to be regarded as an intuitive device employed for pedagogical purposes, not in any sense as a rigorous analysis. However, some further explanation may be helpful. The curve itself is not a social-welfare function in the usual sense; rather it measures in terms of a numeraire (kronor or dollars) the value, summed over all individuals, of the benefits from the output of the activity minus the private *and* net social costs. Thus, for each level of the activity, the height of the curve indicates the *net* benefits (possibly negative) that the activity confers on society. The acceptability constraint indicates that level of the activity which is consistent with the specified minimum standard of environmental quality (e.g., that level of smoke emissions from factories which is sufficiently low to maintain the quality of the air in a particular metropolitan area). There is an ambiguity here in that the levels of several different activities may jointly determine a particular dimension of environmental quality, e.g., the smoke emissions of a number of different industries will determine the quality of the air. In this case, the acceptable level of pollutive emissions for the firm or industry will clearly depend on the levels of emissions of others. If, as we discussed earlier, unit taxes are used to realize the acceptability standards, there will result a least-cost pattern of levels of the relevant externality-generating activities. If we understand the constraint in Fig. 1 to refer to the activity level indicated by this particular solution, then this ambiguity disappears.

is growing evidence that various types of pollutants do in fact have such unfortunate consequences, particularly in areas where they are highly concentrated. [On this see, for instance, Lave & Seskin]. Second, what experience we have had with, for example, the reduction of waste discharges into waterways suggests that processes involving the recycling and reuse of waste materials can frequently be achieved at surprisingly modest cost.[1] In such cases the rationale for the imposition of environmental standards is clear, and it seems to us that the rejection of such crude measures on the grounds that they will probably violate the requirements of optimality may well be considered a kind of perverse perfectionism.

It is interesting in this connection that the pricing and standards approach is not too different in spirit from a number of economic policy measures that are already in operation in other areas. This is significant for our discussion, because it suggests that regulators know how to work with this sort of approach and have managed to live with it elsewhere. Probably the most noteworthy example is the use of fiscal and monetary policy for the realization of macroeconomic objectives. Here, the regulation of the stock of money and the availability of credit along with adjustments in public expenditures and tax rates are often aimed at the achievement of a selected target level of employment or rate of inflation. Wherever prices rise too rapidly or unemployment exceeds an "acceptable" level, monetary and fiscal variables are readjusted in an attempt to "correct" the difficulty. It is noteworthy that this procedure is also similar to the pricing and standards approach in its avoidance of direct controls.

Other examples of this general approach to policy are not hard to find. Policies for the regulation of public-utilities, for instance, typically utilize a variety of standards such as profit-rate ceilings (i.e., "fair rates of return") to judge the acceptability of the behavior of the regulated firm. In the area of public education, one frequently encounters state-imposed standards (e.g., subjects to be taught) for local school districts which are often accompanied by grants of funds to the localities to help insure that public-school programs meet the designated standards. What this suggests is that public administrators are familiar with this general approach to policy and that the implementation of the pricing and standards technique should not involve insurmountable administrative difficulties. For these reasons, the achievement of specified environmental standards through the use of unit taxes (or subsidies) seems to us to possess great promise as a workable method for the control of the quality of the environment.

---

[1] Some interesting discussions of the feasibility of the control of waste emissions into waterways often at low cost are contained in Kneese & Bower. In particular, see their description of the control of water quality in the Ruhr River in Germany.

5. Concluding Remarks

It may be useful in concluding our discussion simply to review the ways in which the pricing and standards approach differs from the standard Pigouvian-prescription for the control of externalities.

(1) Under the Pigouvian technique, unit taxes (or subsidies) are placed on externality-generating activities, with the level of the tax on a particular activity being set equal to the marginal net damage it generates. Such taxes (if they could be determined) would, it is presumed, lead to Pareto-efficient levels of the activities.

(2) In contrast, the pricing and standards approach begins with a predetermined set of standards for environmental quality and then imposes unit taxes (or subsidies) sufficient to achieve these standards. This will not, in general, result in an optimal allocation of resources, but (as is proved formally in the appendix) the procedure does at least represent the least-cost method of realizing the specified standards.

(3) The basic appeal of the pricing and standards approach relative to the Pigouvian prescription lies in its workability. We simply do not, in general, have the information needed to determine the appropriate set of Pigouvian taxes and subsidies. Such information is not, however, necessary for our suggested procedure.

(4) While it makes no pretense of promising anything like an optimal allocation of resources, the pricing and standards technique can, in cases where external effects impose high costs (or benefits), at least offer some assurance of reducing the level of these damages. Moreover, the administrative procedures—the selection of standards and the use of fiscal incentives to realize these standards—implied by this approach are in many ways quite similar to those used in a number of current public programs. This, we think, offers some grounds for optimism as to the practicality of the pricing and standards technique for the control of the quality of the environment.

References

1. Bohm, P.: Pollution, Purification, and the Theory of External Effects. *Swedish Journal of Economics 72*, no. 2, 153–66, 1970.
2. Coase, R.: The Problem of Social Cost. *Journal of Law and Economics 3*, 1–44, 1960.
3. Hart, A.: Anticipations, Business Planning, and the Cycle. *Quarterly Journal of Economics 51*, 273–97, Feb. 1937.
4. Kneese, A. & Bower, B.: *Managing Water Quality: Economics, Technology, In-stitutions.* Baltimore, 1968.
5. Lave, L. & Seskin, E.: Air Pollution and Human Health. *Science 21*, 723–33 Aug. 1970.
6. Portes, R.: The Search for Efficiency in the Precence of Externalities. *Unfashionable Economics: Essays in Honor of Lord Balogh* (ed. P. Streeten), pp. 348–61. London, 1970.
7. Ruff. L.: The Economic Common Sense of Pollution. *The Public Interest*, Spring 1970, 69–85.

APPENDIX

In the text, we argued on a somewhat intuitive level that the appropriate use of unit taxes and subsidies represents the least-cost method of achieving a set of specified standards for environmental quality. In the case of smoke-abatement, for instance, the tax-subsidy approach will automatically generate the cost-minimizing assignment of "reduction quotas" without recourse to involved calculations or enforcement.

The purpose of this appendix is to provide a formal proof of this proposition. More precisely, we will show that, to achieve *any* given vector of final outputs along with the attainment of the specified quality of the environment, the use of unit taxes (or, where appropriate, subsidies) to induce the necessary modification in the market-determined pattern of output will permit the realization of the specified output vector at minimum cost to society.

While this theorem may seem rather obvious (as the intuitive discussion in the text suggests), its proof does point up several interesting properties which are noteworthy. In particular, unlike many of the propositions about prices in welfare analysis, the theorem does not require a world of perfect competition. It applies to pure competitors, monopolists, or oligopolists alike so long as each of the firms involved seeks to minimize the private cost of producing whatever vector of outputs it selects and has no monopsony power (i.e., no influence on the prices of inputs). The firms need not be simple profit-maximizers; they may choose to maximize growth, sales (total revenues), their share of the market, or any combination of these goals (or a variety of other objectives). Since the effective pursuit of these goals typically entails minimizing the cost of whatever outputs are produced, the theorem is still applicable. Finally, we want simply to emphasize that the theorem applies to whatever set of final outputs society should select (either by direction or through the operation of the market). It does not judge the desirability of that particular vector of outputs; it only tells us how to make the necessary adjustments at minimum cost.

We shall proceed initially to derive the first-order conditions for the minimization of the cost of a specified overall reduction in the emission of wastes. We will then show that the independent decisions of cost-minimizing firms subject to the appropriate unit tax on waste emissions will, in fact, satisfy the first-order conditions for overall cost minimization.

Let

x_{iv} represent the quantity of input i used by plant v ($i = 1, ..., n$), ($v = 1, ..., m$),

z_v be the quantities of waste it discharges,

y_v be its output level,

$f_v(x_{1v}, ..., x_{nv}, z_v, y_v) = 0$ be its production function,

p_i be the price of input i, and

k the desired level of $\sum z_v$, the maximum permitted daily discharge of waste.

In this formulation, the value of k is determined by the administrative authority in a manner designed to hold waste emissions in the aggregate to a level consistent with the specified environmental standard (e.g., the sulphuric content of the atmosphere). Note that the level of the firm's waste emissions is treated here as an argument in its production function; to reduce waste discharges while maintaining its level of output, the firm will presumably require the use of additional units of some other inputs (e.g., more labor or capital to recycle the wastes or to dispose of them in an alternative manner).

The problem now becomes that of determining the value of the x's and z's that minimize input cost

$$c = \sum_i \sum_v p_i(x_{iv})$$

subject to the output constraints

$$y_v = y_v^* = \text{constant} \qquad (v = 1, ..., m)$$

and the constraint on the total output of pollutants

$$\sum_v z_v = k.$$

It may appear odd to include as a constraint a vector of given outputs of the firms, since the firms will presumably adjust output levels as well as the pattern of inputs in response to taxes or other restrictions on waste discharges. This vector, however, can be *any* vector of outputs (including that which emerges as a result of independent decisions by the firms). What we determine are first-order conditions for cost-minimization which apply to *any* given vector of outputs no matter how they are arrived at. Using $\lambda_v (v = 1, ..., m)$ and λ as our $(m+1)$ Lagrange multipliers, we obtain the first-order conditions:

$$\left. \begin{aligned} \lambda_v f_{vz} + \lambda &= 0 \qquad (v = 1, ..., m) \\ p_i + \lambda_v f_{vi} &= 0 \qquad (v = 1, ..., m)\,(i = 1, ..., n) \\ y_v &= y_v^* \qquad (v = 1, ..., m) \end{aligned} \right\} \tag{1}$$

where we use the notation $f_{vz} = \partial f_v / \partial z_v$, $f_{vi} = \partial f_v / \partial x_{iv}$.

Now let us see what will happen if the m plants are run by independent managements whose objective is to minimize the cost of whatever outputs their firm produces, and if, instead of the imposition of a fixed ceiling on the emission of pollutants, this emission is taxed at a fixed rate per unit, t. So long as its input prices are fixed, firm v will wish to minimize

$$c = tz_v + \sum_i p_i x_{iv}$$

subject to

$$y_v = y_v^*.$$

Direct differentiation of the m Lagrangian functions for our m firms immediately yields the first-order conditions (1)—the same conditions as before, provided t is set equal to λ. Thus, if we impose a tax rate that achieves the desired reduction in the total emission of pollutants, we have proved that this reduction will satisfy the necessary conditions for the minimization of the program's cost to society.[1]

---

[1] In this case, λ (and hence t) is the shadow price of the pollution constraint. In addition to satisfying these necessary first-order conditions, cost-minimization requires that the production functions possess the usual second-order properties. An interesting treatment of this issue is available in Portes. We should also point out that our proof assumes that the firm takes t as given and beyond its control. Bohm discusses some of the problems that can arise where the firm takes into account the effects of its behavior on the value of t.

PANGLOSS ON POLLUTION

E. J. Mishan

London School of Economics, The American University, London, England

A few weeks ago, a number of us had the pleasure of hearing the celebrated Dr Pangloss being interviewed on the radio. Reproduced below is that part of the interview directed to the current concern with pollution. It is followed by some comments of my own.

E. J. M.

Summary

On being interviewed, Dr Pangloss asserted that optimality is uniquely determined irrespective of legal or distributional considerations; moreover, that pollution above a theoretical optimum could be justified economically in so far as the heavy transactions costs of correcting the existing suboptimal position are real costs and exceed the optimality gains.

A change to anti-pollution laws, however, not only alters the optimal position but reverses the role of the transactions–cost barrier, the result being "too little" pollution. Since transactions costs are inevitable in any change to optimum, which laws should society choose? The case for a change to anti-pollution laws rests on equity and allocative merit, with particular attention being paid to the external diseconomies of increasing risk.

I

Commentator: Dr Pangloss, I understand you to say that there is no pollution problem.

Pangloss: That is correct. By any standards pollution must be vastly less than it was a hundred years ago. In fact, the great triumph of modern civilization has been the reduction of pollution.

C.: Yet there does seem to be a growing concern by scientists and by men of affairs, including some economists, over the quality of the air, the disposal of garbage, the pouring of effluents into streams, lakes, and into the seas ....

P.: But this concern is only one more instance of the more articulate and professional groups trying to foist their own tastes and systems of values onto the rest of society. It is like the demand for subsidizing radio stations to broadcast classical music, opera, or Shakespearean plays. Thus, the demand for clean air is primarily an upper income demand—the high-income people want to get the low-income people to pay for something that the high-income people value.

C.: Your are saying that poorer people have no interest in such refinements?

P.: In all my experience, and in all my reading of history, I have never yet come across any unambiguous evidence of a demand among poor people for quiet or clean air. If anything, the evidence points the other way; the poorer the people, the more they tend to settle in the dirtiest and noisiest parts of the city.

bollocks !!

C.: But if there *were* a pollution problem, Dr Pangloss, what sort of a solution should we seek? Is there not a case for government intervention?

P. By no means. People can act in their own interests far better than other people can act for them.

C.: But what can a person do who lives in a polluted urban area?

P.: Clearly, if he does not like it he can move elsewhere—at least he can do so in a free society. However, you will observe that, on the whole, people move from the clear clean countryside to the polluted cities. We must therefore infer that the advantages of the city outweigh the disadvantages.

C.: But now take a simple individual instance: suppose the person next door is driving me into a frenzy by incessantly working his new motorized lawn mower, or his new motorized hedge-trimmer, or his new soil-turner, or something new and motorized, what can I do to preserve my sanity?

P.: Why, you can bribe him to stop, or to fit silencers to his motors.

C.: But is that fair?

P.: Perfectly fair. Why should you interfere with his enjoyment without compensating him? Besides which, if you can bribe him to stop, both he and you are made better off than before. Such a solution is clearly ideal .... .

C.: But suppose I cannot afford to bribe him?

P.: Why, in that case the existing situation is already ideal or—as we economists say—is 'optimal'. For clearly it is then no longer possible to make both of you better off by stopping your neighbour's gardening activity.

C.: And does the same sort of argument apply also to aircraft noise?

P.: Why, surely. If there are enough people suffering from aircraft noise, they can combine their resources and bribe the airport to move elsewhere.

C.: But organizing a large number of people to translate protests into financial payments, and to engage in litigation, and so on, would be so risky, troublesome, and perhaps so expensive, as to make your proposition quite impractical.

P.: Ah, but that only goes to show that the existing arrangement, the noisy airport, is actually an optimal position.

C.: How do you figure that out?

P.: Simple. In order to show that the existing, noisy airport plus airline services is *not* optimal, it is necessary to show that the gains from moving the airport elsewhere exceed the costs. These costs not only include the losses to be borne by the airport (or airlines) but also those costs of organizing the protest, and the costs of negotiation and litigation of which you have spoken.

But all these costs taken together are (as you say) too great; the gains from moving the airport cannot exceed them. Therefore the existing situation is, as I observed, already optimal.

C.: You would say that of all existing noisy airports?

P.: From the observation that they have every intention of continuing business there, we cannot but infer that the situation is optimal.

C.: Then it seems that this is the best of all possible worlds after all?

P.: Good as things are, they would be a lot better if government intervention were less.

II

The first thing to be said of Dr Pangloss's arguments is that they are perfectly valid in a Panglossian universe—one in which a variety of pollutants are concentrated only in certain areas, say urban areas; in which accurate and relevant knowledge is available to all without cost; in which the question of justice is irrelevant, and in which there are no future generations to consider. Once we move away from the Panglossian universe toward the world we live in, however, doubts begin to assail us. Let us explore the analytic genesis of these doubts.

(1) If you cannot bribe your neighbour to stop his nuisance activity then, according to Pangloss, the existing situation is optimal. This rather hard doctrine deserves more careful examination. An imaginary case history lends itself to this purpose.

Jim lives next door to George on the outskirts of Pleasantville. George, who has recently bought himself the new "Compleat Gardener" outfit, has become such an enthusiast that Jim's week-ends have become nightmares. Having accidentally tuned in to the interview with Dr Pangloss, it occurs to Jim to approach George with an offer. After exchanging a few rude words, some haggling takes place, and it emerges that Jim is prepared to pay, at most, $60 each week if George will desist from his new hobby during week-ends. George, however, refuses to accept a cent less than $70. Recalling Dr Pangloss's fateful words, Jim gloomily concludes that the existing intolerable situation is ideal and resigns himself to an early grave.

Now it so happens, about a year later, a new government comes into power, headed by men who know not Pangloss. Among other radical measures they alter the existing law so that no man may use a motorized garden implement without express permission of his neighbour. Pangloss at once goes on record as averring that such measures are futile. They cause merely distributional changes. The existing real situation, however, would, he predicted, remain unchanged: there would be no less gardening and no less noise.

Jim, of course, is overjoyed with the "merely distributional" effects. As he anticipates, George comes round soon after and politely enquires whether Jim will accept $60 a week to put up with the usual week-end racket. Jim assures him that he cannot take a penny less than \$70, while George swears he cannot afford to pay more than \$60. Each accuses the other of deceit and obstinacy. But no agreement is reached, and there are no more week-end gardening sessions for George.

If Jim and George are both telling the truth then, indeed for the original noise-permissive law, George's week-end gardening activity is optimal whereas with the new anti-noise law the quiet week-end is optimal, and we must infer that— contrary to Dr Pangloss's statement—the optimal outcome does, after all, depend upon the law. But can they both be telling the truth?

Consider, first, Jim's response. He would pay up to \$60 to stop the noise, but he would want \$70, at least, in order to put up with it. Is this consistent? It is certainly consistent with current price theory. Call the original noisy situation I, and the no-noise situation II. Jim's welfare in the initial I situation is indicated by W_1; his welfare in the initial II situation by W_2[1]—with W_2 being greater than W_1. If his "income" effect (or, more accurately, his welfare effect) is normal, or positive, then at the W_2 level Jim is willing to pay more money for quiet, or will require more money to forgo it, (say \$70) than he is at the W_1 level (say \$60). The same logic[2] applies also to George—except that his welfare is higher in the initial I situation than in the initial II situation.

It follows, therefore that under the noise-permissive law Jim is unable to bribe George to stop his motorized gardening, the existing noisy weekends outcome being optimal. It follows also that under the noise-prohibitive law, George is unable to bribe Jim to put up with the din, and quiet weekends are the optimal outcome.[3]

(2) Turning to the airport example (and from now on ignoring welfare effects), if it were somehow known that the victims of aircraft noise were together able to pay up to \$10 million and the losses incurred by the airline interests in closing the airport, or moving it elsewhere, were \$8 million, there would appear to be a clear case for moving it elsewhere. But, as Dr Pangloss sagely remarked, if the costs of discovering these facts, and of negotiating and organizing the transfer of compensatory payments, were in excess of \$2 million then the existing

---

[1] I say the *initial* I situation to refer to the respective welfare levels of Jim and George to start with. For they could have begun in situation II and by mutual agreement returned to the I situation at somewhat different levels of welfare. Similar remarks apply to the initial II situation.

[2] In more extreme cases, the budget restriction places a limit on the amount a man can pay whereas there is no such limit on the amount he can receive. For example, in order to pay for an operation that will save the life of his only child a man cannot raise more than \$100 000. But the minimum sum he would accept to go without the operation, and thus risk the loss of his child's life, might be infinite. Perhaps no amount of worldly goods could restore his level of wellbeing after losing his child.

[3] Put more formally, if the payment of a compensatory sum is positive, the receipt of a compensatory sum negative, the algebraic sum of the compensating variations over all persons affected with respect to any specific change is positive if the change entails a potential Pareto improvement. Since, in the above instances, the algebraic sum $(-70 + 60 = -10)$ is negative, a potential Pareto improvement is not possible. The existing situation is then optimal in each case, as stated.

unchecked-aircraft situation must, after all, be optimum. Indeed, when pressed on the matter later, Dr Pangloss extended such auxiliary costs to the time and effort spent by economists in devising methods (other than direct negotiations) of tackling spillover problems, to research in calculating and frequently revising sets of (ideal) taxes or regulations for polluting industries, to administrative and inspection costs, to the waste of entrepreneurial effort in evading such taxes or regulations, and so on. He went on to conjecture that these so-called transactions costs, to say nothing of frictional costs, were far heavier than would-be reformers thought. In the cloud-cuckoo land of welfare theorists there would, he said, appear to be unlimited opportunities for Pareto improvements. But this appears to be so, simply because they never pause to consider the magnitude of these real transactions costs which are incurred in initiating, effecting, and maintaining these imaginary improvements. The mere fact, concluded Dr Pangloss, that despite such apparent opportunities no reductions in pollutants are taking place—except those arising from unwarranted interference by government officials—is itself substantial evidence that existing arrangements are optimal in the over-all sense.

Dr Pangloss is again right, but he does not realize that the argument is double-edged. To show this, let us start with laws that permit unchecked pollution. The existing market outputs, M_1, M_2, M_3, ..., of highly competitive polluting industries, Z_1, Z_2, Z_3, ..., are in general larger than the set of costless optimal outputs, Q_1, Q_2, Q_3, ..., (where some Q_i can remain unchanged at their existing market outputs, or else become zero).[1] If, on the other hand, we start off with universal anti-pollution laws, to the effect that no polluting activity is permitted without the express consent of all parties affected, the initial market outputs, M'_1, M'_2, M'_3, ..., of these polluting industries, Z_1, Z_2, Z_3, ..., would be either zero, or smaller, or no greater than the corresponding market outputs, M_1, M_2, M_3, ..., produced under the permissive law. Again, however, in the absence of all transactions costs, these initial market outputs, M'_1, M'_2, M'_3, ..., would be expended to these hypothetical optimal outputs, Q_1, Q_2, Q_3 ... .

Thus, under pollutant-permissive laws the market outputs M_i are truly optimal in the over-all sense that includes transactions costs, and always larger than the M'_i outputs that are also optimal in the over-all sense under the anti-pollution laws. In general, that is, $0 \leqslant M'_i \leqslant Q_i \leqslant M_i$.[2]

---

[1] For some Q_i, pollution may be invariant to output, or very nearly.

[2] It has been alleged by Dr Mansholt that it is politically more feasible to introduce anti-pollution legislation into a country as part of an international, or at least multi-national, agreement, since otherwise a country might find itself at a disadvantage in competitive world markets. Now to the extent a product inflicting (uncosted) spillovers on the home country is being exported, the foreigner is effectively subsidized by the home country. Requiring spillover costs to be covered (or imposing appropriate taxes) is then equivalent to removing the subsidy. The welfare effects from trade of removing an export subsidy are, of course, no more certain than those of offering an export subsidy. Whatever the actual outcome is, however, there is nothing to prevent the home country, if it wishes, to continue to subsidize the same export quantities as before, being, then, no worse off in

The thrust of the above argument is that transactions costs act as a formidable cost-barrier, protecting the existing market solution against hypothetical Pareto improvements. But which side of the barrier society starts from is not determined by economic mechanisms. If society chooses its laws so as to place itself on the pollutant-permissive side of the cost barrier, the over-all optimal solution (which may coincide with the existing market solution) will involve more pollution than a costless ideal solution, and more pollution still than the over-all optimal solution that emerges if society, instead, chooses its laws as to place itself on the other side of the cost barrier.

In the particular case of antiaircraft-noise legislation, this means that, even if it were the case that aircraft interests could more than compensate the victims of aircraft noise, the costs of negotiation or working out sets of ideal taxes may be so heavy that the over-all optimum outcome may be zero aircraft flights.

(3) It will be recalled that when it was put to Dr Pangloss that having to bribe the polluter might be unfair, it was roundly denied. Asked to elaborate on this judgement later on, he argued that the conflict of interest was symmetric in all respects. He quoted the well-known case of two small businesses sharing a single building. Mr Hyde, who turned out leather soles, used machinery, the noise and vibration from which disturbed the work of Mr Stone, the diamong setter. Who should compensate whom? asked Dr Pangloss. Which of them should pay for the anti-noise devices? If, as alleged, Mr Hyde's activities cause Mr Stone to lose money, it is also true that compensating Mr Stone will cause Mr Hyde to lose money. Clearly, concluded Dr Pangloss, there is no equitable way of deciding which of them shall pay for the anti-noise device.

The fact that a conflict of interest exists does not, however, always preclude judgement about culpability. If a burglar breaks into my house in order to steal my wife's jewelry, and I try to prevent him, there is a clear conflict of interest. For that matter, given the burglar's intention and his appraisal of the effort and risks, an effective bribe to desist may be an over-all Pareto optimal solution. Yet it would be hard to conceive of any organized society having difficulty in deciding the rights of such a conflict.

Certainly if we restrict ourselves to pollution falling on the public, it is not difficult to decide the case in equity. According to the libertarian philosophy of John Stuart Mills, the freedom of a man to do as he wishes must not go so far as to infringe the freedom of his fellows. If George, in his addiction to motorized gardening, shatters Jim's week-ends, the case in equity is not sym-

---

respect of trade than it was before—save that now it pays the subsidy overtly rather than, as before, covertly (through ignoring the costs of pollution).

The extent to which unheeded spillover effects do in fact subsidize exports is, of course, an empirical question. There may be quite a range of uncorrected spillovers, in particular localised spillovers, that have negligible or zero effects in reducing the prices of the home country's exports.

metrical. The freedom that George would claim certainly harms Jim. In contrast, Jim's claim for quiet and clear air does not, of itself, harm George.[1]

On the grounds of equity, if nothing else, there is then a case for effective anti-disamenity legislation, prohibiting any form of pollution in the absence of mutual agreement. Some will point to the costs of implementing such legislation, but these have now to be borne by the pollutors[2] who also, incidentally, have an incentive to direct ingenuity and research into discovering cheaper ways of securing agreement among all members of society.

(4) Dr Pangloss was in the true liberal tradition in asserting that each person should be allowed to judge of his own interest. In the pollution context, he was, of course, thinking of the liberty of the victim of pollution to offer any bribe he wishes to the pollutor, of his liberty to move himself and his family anywhere he wishes at his own cost—assuming there are still unpolluted areas to move into. If Dr Pangloss could be persuaded that new laws requiring that pollution victims be compensated were no less tolerable than existing laws, it might be thought that an accord could be reached with him. But not quite. Since in fact people do *not* always know their own interests best, there may be occasions in which the government should disallow arrangements freely entered into to the greater interests of society.

As a general rule, it may be opined that, as a matter of political expediency at least, governments should act as if people knew their own interests best. But this maxim springs from the view that adults have enough experience of the world, or may come to acquire it over time as to be able to act on their own behalves without substantial injury to themselves or others. In the world we are now living in, however, this view is no longer tenable. For the gathering pace of technology acts to extend the time lag between the immediate (commercial) exploitation of new products and processes, on the one hand, and the general recognition, on the other, of their short- and long-term genetical and ecological effects.

The risks arising from insufficient knowledge of the effects over time of

---

[1] This distinction between the 'active' and 'passive' agent in a conflict of interest is tenable even though tenuous cases may be resolved only by recourse to an ethical consensus. If the mere sight of a person, or the mere knowledge of his existence or behaviour, offends me, it is not likely that society would support my claim to damages. Apart from determining the validity of this distinction in fine cases, the ethical consensus may qualify the concept of culpability by reference to the context in which the alleged offence occurs. There would be little sympathy with a man who complained of the yelling by fans at a football match, or of the noise and smoke inside a pub, or of indecent exposure in a brothel. Such features are customarily associated with such places, and they can obviously and easily be avoided. The case in equity is different if at the time of entering into a heavy commitment a person has a reasonable expectation of the continuance of certain amenities. Indeed his plight will command sympathy even though no such anticipations can be harboured, if, at the same time, feasible alternatives no longer exist.

[2] I do not envisage extending the law of torts in this respect so much as making it a grave offence against the law for a person or firm to engage in any of a range of activities without official permission, which permission is to be annually renewed, and withheld unless all 'reasonable' claims are satisfied, or else unless such technology is employed that no 'effective' pollution is generated. (The quoted words indicate that more precise criteria must depend upon further debate, and on administrative and legal experience.)

any single innovation may well be thought small. The genetical effects of Thalidomide wrought individual tragedies but they were discovered before humankind was endangered by its universal adoption as a powerful sedative. The ecological consequences of widespread application of DDT, and other chemical pesticides, are coming to light more tardily, but they are not expected to endanger the human species. But even if the risk associated with any single innovation were small—and there are not many who believe they are—as the numbers of such innovations increase and spread over the globe, the chance of some uncontrollable epidemic, or ecological disaster, occurring is sure to grow rapidly—unless controls are imposed that are far more stringent than those currently contemplated in the West. In the interests of human survival, then, there is a strong case for some curbing of free enterprise and consumers' sovereignty; a case for stronger public control over commercial exploitation, or application, of new processes and over the marketing of chemical and other synthetic products.

Another consideration reinforcing the case for occasional government intervention, in despite of freely negotiated agreements under any pollution legislation, arises if we expect humankind to survive. Successive future generations can suffer from the effects of any prolonged or virtually irrevocable spillover damage caused by existing generations—one thinks in this connection of the effects on natural beauty all over the world of the mushroom growth of package tourism. But since they have no representation in decisions taken today, any apparent optimal damage based on a current consensus will be greater than one that would be valid over time.

III

Let me conclude. Dr Pangloss has many followers. For nothing yields more satisfaction to an economist than to dissect the schemes of ardent reformers, and to demonstrate that any interference with the free pay of self-interest must invariably make matters worse. I will therefore summarize my main points, and leave it to the Panglossians to score again, if they can.

1. Both because of welfare effects and because of the magnitude of transactions costs the over-all optimal outputs of polluting activities under pollution-permitting laws can be much larger than their over-all optimal outputs under pollution-prohibiting laws.

2. In the case of environmental pollution, there appears to be a *prima facie* case in equity against the pollutor *vis a vis* the non-pollutor.

3. For a range of economic activities, there can be a case for government controls even though a consensus is reached under any kind of pollution laws (*a*) because some adverse spillovers created today will fall also on future generations, and (*b*) because our knowledge of a growing number of new processes and synthetic products is so limited that, unless we develop institutions to control their use and sale, we run an increasing risk of universal catastrophe.

Chris Bradshaw
is my
Hero.

Part III
MEASURING THE VALUE OF ENVIRONMENTAL DEGRADATION

A PLANNING APPROACH TO THE PUBLIC GOOD PROBLEM

*E. Malinvaud**

Institut National de la Statistique et des Etudes Economiques, Paris, France

Summary

The theory of planning should give new insight into the classical problem of how to achieve an optimal provision of public goods. The point is exhibited by a diagrammatic study of the simple model in which there are just two consumers and two commodities, one private, one public. Besides the procedure proposed by E. Lindahl, two others are discussed; the first uses tax indicators, the second quantity indicators. Consideration is given to three requirements: convergence to an optimum, equitable treatment of the two consumers, incentives for correct revelation of preferences.

1. Planning for the Provision of Public Goods

The recent development in the analysis of collective consumption has interest for both aspects of economic theory. On the one hand it aims at being "positive", i.e. at explaining how the provision for public goods is actually decided in our societies. On the other hand it is also "normative" when it suggests ways for a better organization of social decisions in such matters. The two aspects are often simultaneously involved, the border line being more difficult to draw in this subject than in many others. Indeed by its very nature public consumption requires that some direct agreement be reached between those taking part in the government of the community. We do not find in this field the clear distinction between the positive study of an equilibrium resulting from individual behavior and the normative study of an optimum program for the society as a whole.

It seems, however, that substantial progress in our understanding of the subject may come from a deliberate attempt at following the normative line to its extreme end. One should not stop when a clear definition of an optimum has been achieved and when some duality property has shown the existence of prices and tax rates with respect to which an optimum appears as some kind of "pseudo-equilibrium".[1] One should, for collective consumption still more than for other questions of welfare economics, inquire about the social decision

---

* This work was presented for discussion during a very stimulating visit I made to Stockholm, Lund and Copenhagen in October 1969. I benefited very much from the suggestions I received then.
[1] See P. A. Samuelson [7].

procedures that are likely to bring about an optimum. If we define the "theory of planning" as studying these procedures, and as paying particular attention to the fact that information on wants and technological possibilities is a priori decentralized, then it becomes clear that our understanding of the public good problem may gain substantially from the theory of planning.

A process through which an optimum program is determined can be viewed either as an iterative exchange of information between agents and a central administrative body or as a bargaining discussion between representatives of various social groups, a discussion that follows some institutional rules. A theoretical study of the properties of alternative processes is likely to exhibit their respective qualities or deficiencies and therefore to help for the organization of the public economy.

In order to substantiate these points, I shall consider here the simplest of all models concerning collective consumption, a model discussed by E. Lindahl [3] and many others. Generalizations are possible.[1] But looking at the simplest case will reveal in full clarity some important elements of any theoretical study concerning decision procedures for the choice of a public consumption program.

2. The Model

Let there be only two "consumers", individuals or social groups, $(i = 1, 2)$ and two commodities: the public good whose quantity will be denoted z and the private good consumed in quantities x_1 and x_2 by the two consumers. Let us assume that the feasibility constraints are simply that z, x_1 and x_2 be non-negative and that the sum of these three quantities be a given number ω (the rate of substitution betwen the private and the public good is fixed and taken equal to one).

A feasible "program" will therefore be defined by three non-negative values z, x_1, x_2 that fulfil the equality:

$$x_1 + x_2 + z = \omega. \tag{1}$$

In conformity with the usual approach to the problem let us assume that no external effect exists in the consumption of the private good. Individual i will then rank the programs by considering only the two values (x_i, z) that concern him. His preference ordering may conveniently be represented by a utility function, i.e. by a function $U_i(x_i, z)$ of the two arguments x_i and z.

The most common formalization of equilibrium stipulates that consumer i has an income y_i, but that a part $t_i z$ of the latter is raised as a tax for financing collective consumption. The disposable income $y_i - t_i z$ is used on the private good whose price can be taken as 1. Hence the budget constraints are:

$$x_i + t_i z = y \quad i = 1, 2. \tag{2}$$

[1] See in particular E. Malinvaud [5].

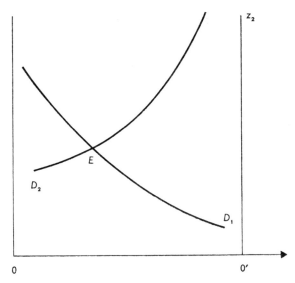

Fig. 1

In view of (1) the price of the collective good is also equal to 1, so that the number t_i is the proportional share of consumer i in the tax bill. Hence

$$t_1 + t_2 = 1. \tag{3}$$

Summation of the two equations (2) then imply:

$$y_1 + y_2 = \omega. \tag{4}$$

In order to study the determination of an equilibrium, E. Lindahl [3] considered the "demand" for the public good, i.e. the value z_i of z that consumer i would prefer if he could choose it freely, his income y_i and his share t_i being imposed on him. The vector (x_i, z_i) therefore maximizes U_i under the constraint (2). For a given y_i this demand may be considered as a function of t_i. Hence Lindahl drew a demand curve D_1 for the first consumer with t_1 on the horizontal axis and the corresponding z_1 on the vertical axis (see Fig. 1). Due to equality (3) the same figure can be used for the representation of the demand curve of the second consumer if a new vertical axis is drawn for z_2 from the point $0'$ on the horizontal line where $t_1 = 1$ (the share t_2 is then plotted from $0'$ to the left). The point E where the two curves intersect each other defines an equilibrium, since $z_1 = z_2$ and the summation of the two budget constraints implies (1).

As is wellknown, demand curves are not very convenient for studying optimality questions. I shall therefore use a different geometrical representation that is reminiscent of the classical Edgeworth box in which the indifference curves of the two traders directly appear.[1]

---

[1] Such a diagram was used in research papers by S. Ch. Kolm.

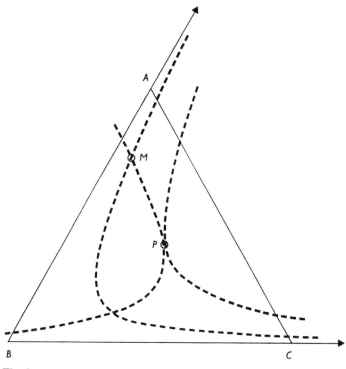

Fig. 2

The feasible programs may be represented as the points of an equilateral triangle ABC of vertical height ω (see Fig. 2). The vertical distance from a point M to the horizontal side BC is taken as z, the distance from M to side AB as x_1, finally the distance from M to AC as x_2 (geometry tells us that the sum of these three distances does not depend on the exact location of M and is therefore equal to ω). In this triangle we may represent the indifference curves of the two consumers, i.e. the contour lines of $U_1(x_1, z)$ and $U_2(x_2, z)$ respectively. For instance the indifference curves of the first consumer may be plotted by reference to the two axis BC and BA; except for inclination, they have the traditional form (two curves are drawn with their concavity to the right). Similarly the indifference curves of the second consumer, plotted with reference to the axis CB and CA, turn their concavity to the left (one curve is drawn).

The two budget lines (2) could also be drawn in this triangle. As long as (3) and (4) hold, they would be represented by *the same* straight line (they would be straight lines because the mapping from the triangular coordinates to the common cartesian coordinates is linear; they would coïncide with each other because $x_1 + t_1 z - y_1 = (\omega - z - x_2) + (1 - t_2)z - (\omega - y_2) = -x_2 - t_2 z + y_2)$. The line would intersect the horizontal side at the point H whose distance from side AB

is y_1 and from B is $2y_1/\sqrt{3}$. It would pass to the left or to the right of A depending on whether t_1 would exceed or not y_1/ω. (Budget lines will be drawn in Figs. 3 and 4.)

On this representation we immediately see which programs are Pareto optimal. A program like M could be improved for both consumers by reduction of the collective consumption z. On the contrary a program like P, where the indifference curves of the two consumers are tangent to each other, cannot be improved and is therefore Pareto optimal. Like in the classical Edgeworth box, there will usually be a line of Pareto optimal programs. Moving along this line from left to right one will find programs that are more and more favorable to consumer 1, but more and more unfavorable to consumer 2.

The planning problem occurs when a particular program must be selected in a situation where information is not centralized from the beginning. Such being the elements of our simple model, we must consider the case in which the central institution (we shall call it the "board") does not know the indifference curves of the two individuals, each one of them knowing only his own preferences. We must study procedures for the exchange of information and the progressive determination of the program to be selected.

According to an approach that I defined elsewhere [4], I shall formalize the exchange of information as taking place at successive times or "stages" (with $s = 1, 2, \ldots S$). The exchange at stage s will consist of (i) "indicators" issued by the board and sent to the two individuals and (ii) "proposals" formed by these individuals to the board. At the last stage S the board selects one program (we may call it the "plan" whether it is to be implemented directly or simply suggested for agreement between the two individuals).

3. Lindahl's Solution

Although it was intended at giving a "positive solution", the classical article of E. Lindahl [3] may be considered as a contribution to the planning approach. Stated in my own words it would be presented as follows.

Each consumer has a fixed income y_i, condition (4) being fulfilled. The board uses as indicators the contribution rates t_i and chooses them in such a way as to satisfy (3); the proposals of the individuals are their demands for private and public consumption. At stage s consumer i, considering the rate t_i^s that was indicated to him, determines his demands x_i^s and z_i^s as the components of the vector maximizing his utility $U_i(x_i, z)$ under the budget constraint

$$x_i + t_i^s z = y_i. \tag{5}$$

In order to define completely the procedure, it is sufficient to say how the rates are revised from one stage to the next and how the "plan" is determined. (The initial rates t_1^1 and t_2^1 may be considered as defined a priori according to some rule intended at fairness or equity, for instance as being equal to $\frac{1}{2}$.)

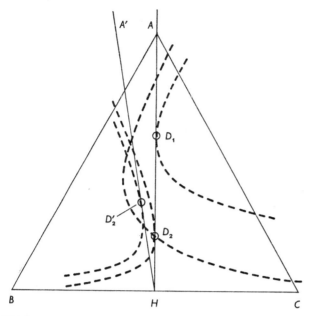

Fig. 3

Lindahl's solution is then based on two principles: (i) When the two proposals z_1^s and z_2^s do not coïncide, the rate should be raised for the individual asking for the higher quantity of collective consumption; it should correspondingly be lowered for the other individual. (ii) The individual that sees his rate being thus raised will agree as long as the lowering of the rate of his companion will induce from the latter a demand that he himself considers as being an improvement.

In order to understand better the situation, let us consider Fig. 3 where the vertical line AH is intended to represent the initial budget constraints (with $y_1 = y_2 = \omega/2$ and $t_1^1 = t_2^1 = \frac{1}{2}$). The demands of the two individuals are represented by the points D_1 and D_2 where the budget line is tangent to the respective indifference curves. If these two points would coincide, they would define a feasible program that would be both an equilibrium and Pareto optimal. No revision would be required. But in Fig. 3 individual 1 is asking for much more collective consumption than individual 2 does.

What will then happen if the rate t_1 is increased for individual 1? The budget line will rotate around H to $A'H$. This change will induce a revision of the demands. In particular it is likely to lead to a higher demand for collective consumption from the second individual who benefits from a reduction of his contribution rate. This higher demand may be favored by the first individual, even though he must contribute a larger share to its financing. In Fig. 3 the new point D_2' is indeed better than D_2 for individual 1 as well as for individual 2.

Lindahl assumes that, when the two consumers do not agree, the collective consumption will be fixed at the minimum of the two demands. With the budget line AH of Fig. 3 the actual program will therefore be represented by D_2. Facing such a situation, individual 1 finds that it is in his interest to accept an increase of his rate t_1.

But such an approach will not lead to an optimum. At some stage in the revision of his rate t_1, individual 1 will see that a further increase would reduce his individual consumption too much to compensate for the increase in collective consumption. Indeed the indifference curves of Fig. 3 could have been drawn in such a way that D_2' would have given to 1 a lower level of utility than he had with D_2. Such a situation will necessarily occur in a neighborhood of an optimum program because the difference between the utility levels achieved by consumer 1 under his own demand D_1' and under the demand D_2' of the other will be of a second order of smallness with respect to the distance $D_1'D_2'$.

The point was clearly made by Lindahl when he presented his "positive solution". The same procedure was recently proposed by L. Johansen [2] as a way of achieving an optimal program by some planning process. The second principle posed by Lindahl' must then be removed: in our case individual 1 must be required to report his demands under increasing tax shares t_1, even beyond the point where he would benefit from the revision of the demand of individual 2.

When it is thus amended the procedure suggested by Lindahl exhibits in full clarity one particularity that was already present in its original version: it is not satisfactory from the point of view of equity. It is unfair to the individual who most need collective consumption: the discussion leading to the revision takes as a starting point the demands of the other individual and almost completely neglects his own. We certainly want a planning procedure to exhibit more neutrality. But the two deficiencies of the argument reproduced in this section can be remedied if we change a little the procedure.

4. A Procedure with Mutual Concessions

Assume first that, when the two individuals do not agree, the collective consumption will be fixed at an average of the two quantities that are demanded respectively by 1 and 2. In Fig. 4, where for clarity the triangle is not drawn, the point P middle of the segment $D_1 D_2$ represents the program that would be selected if there were no revision.

Assume also that the revisions of the contributions are organized in such a way that the budget line rotates not around H but around the middle point P. Then the two individuals are treated equitably. Moreover it is intuitively clear in Fig. 4 that the discrepancies between the demands will tend to be resolved if the revision of the contributions is made in the proper direction.

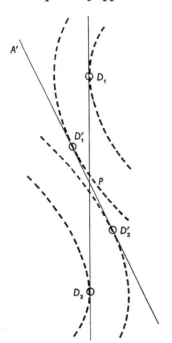

Fig. 4

Let us therefore see what is meant by a shift of the budget line from AH. to $A'P$. We easily write the components of the program P:

$$\hat{z} = \tfrac{1}{2}(z_1 + z_2) \tag{6}$$

$$\hat{x}_i - x_i = -t_i(\hat{z} - \hat{z}_i) \qquad i = 1, 2 \tag{7}$$

the equalities (7) following from the fact that P as well as the demands of the individuals satisfy the budget constraints (the index s is omitted for simplicity). Since the first individual asks for more collective consumption than is provided by P, it is natural that t_1 be made greater than t_2, or more generally that t_i be increased for the individual that is asking for more than $\hat{z}$. We can consider for instance the rule:

$$t_i^{s+1} - t_i^s = b(z_i^s - \hat{z}^s) \qquad i = 1, 2 \tag{8}$$

where b is a fixed positive coefficient defining the speed of the adjustment. If we want to rotate the budget line around P and not around H, we must revise the incomes of the two individuals or, equivalently, impose transfers among them. Let us write the sum of the income and the net transfer received as $y_i - T_i$ or the total net tax levied on consumer i as $t_i z + T_i$. At the beginning T_1 and T_2 will be taken equal to zero. Moreover, in order to preserve equality (4) we must impose:

$$T_1 + T_2 = 0, \tag{9}$$

which is quite satisfactory for a transfer payment. We want the budget line:

$$x_i + t_i^{s+1} z = y_i - T_i^{s+1} \qquad i = 1, 2 \tag{10}$$

to contain the program P, after as well as before the revision, hence we must have:

$$T_i^{s+1} - T_i^s = -(t_i^{s+1} - t_i^s)\hat{z}^s. \tag{11}$$

The transfer T will therefore be revised in favor of the individual who is asking for the larger collective consumption and therefore has to accept an increase in the rate t_i concerning him.

It is intuitively clear, and may be proved formally in a more general model,[1] that the procedure so defined has two interesting properties if the revisions are made by small enough steps.

(i) It implies mutual concessions from all the individuals in the sense that, from one iteration s to the next $s+1$, the utility U_i of the consumption vector (x_i, z_i) selected by individual i decreases:

$$U_i(x_i^{s+1}, z_i^{s+1}) < U_i(x_i^s, z_i^s) \tag{12}$$

for all i; all individuals share the burden of improving the consistency of their demands.

(ii) It converges to an optimal program since the plan $(\hat{z}^s, \hat{x}_1^s, \hat{x}_2^s)$ progressively approaches the line of Pareto optima as s increases; in other words, the revisions will continue up to an agreement on the program and the corresponding tax parameters t_i and T_i.

5. A Procedure for the Improvement of a Feasible Program

Considered as planning procedures the two processes defined above use as "indicators" the tax shares t_i, which play a role similar to that of prices in the traditional theory. Correspondingly the "proposals" are quantities demanded by the consumers. It is of course possible to reverse the roles of prices and quantities, the latter appearing as indicators and the former as proposals. It even turns out that the resulting procedure looks more interesting than the two preceding ones.

Let us therefore assume that the board selects at each stage a feasible program (z^s, x_1^s, x_2^s), i.e. three quantities fulfilling equation (1) and defining a point within the triangle of our graphical representation. Each individual, to whom the program is announced, will not be surprised if he is asked to report how much he would contribute at most, by restricting his private consumption x_i^s, in order to see the collective consumption z^s being increased by one unit (this latter being assumed small). More precisely his proposal will

---

[1] See E. Malinvaud [5].

have to be his marginal rate of substitution between collective and private consumptions:

$$\pi_i^s = \frac{U'_{iz}}{U'_{ix}} \tag{13}$$

U'_{ix} and U'_{iz} being the two derivatives of the utility function U_i with respect to x_i and z, both being evaluated at (x_i^s, z^s). The π_i^s have also been called "marginal willingness to pay" in the theory of public goods.

The board receiving the two proposals π_1^s and π_2^s will see whether an increase or a decrease in z is required. It will raise or lower the collective consumption depending on whether the sum of the marginal rates π_i^s exceeds 1, which is the marginal rate of substitution from the point of view of feasibility. A precise rule may for instance be:

$$z^{s+1} - z^s = b(\pi_1^s + \pi_2^s - 1), \tag{14}$$

b being a fixed positive number.

The simultaneous revision of the private consumptions must be such that each individual does not contribute more than in proportion to his marginal willingness to pay. But this may always be achieved. For instance the following rule may be applied:

$$x_1^{s+1} - x_1^s = \tfrac{1}{2}(\pi_2^s - \pi_1^s - 1)(z^{s+1} - z^s)$$

$$x_2^{s+1} - x_2^s = \tfrac{1}{2}(\pi_1^s - \pi_2^s - 1)(z^{s+1} - z^s) \tag{15}$$

Considering the first of these two equations we see that, if z^{s+1} exceeds z^s, $\pi_1^s + \pi_2^s - 1$ is positive, then $\tfrac{1}{2}(\pi_2^s - \pi_1^s - 1)$ is greater than $-\pi_1^s$, hence $x_1^{s+1} - x_1^s$ greater than $-\pi_1^s(z^{s+1} - z^s)$: the reduction of the private consumption of the first individual is smaller than the one he was ready to accept. Conversely, if the collective consumption is reduced, the increase in the private consumption of each individual is larger than the amount required for compensating him for this reduction. Summation of the two equations (15) finally shows that the new program $(z^{s+1}, x_1^{s+1}, x_2^{s+1})$ is feasible as well as the preceding one.

In the graphical representation the above considerations mean that the point P^{s+1} representing the new program will be located within the region that is defined by the two indifference curves passing through P^s (see Fig. 5).[1]

---

[1] A geometrical argument can be given. Consider on Fig. 6 the equilateral triangle P^sRS with upper vertex at P^s and height equal to 1. (Since the unit is assumed to be small, the scale of Fig. 6 must be seen as greatly magnified with respect to that of Fig. 5.) Let P^sQ_1 be the tangent to the individual 1 indifference curve that passes at P^s, the point Q_1 being on the horisontal side of the triangle (Q_1 might be on the right of S). Let K_1 be the projection of Q_1 on P^sR. Then the length of Q_1K_1 is equal to the maximum amount of private consumption that individual 1 is ready to give up in order to see collective consumption increase by one unit; the length of Q_1K_1 is therefore equal to π_1^s. Similarly we may define P^sQ_2 and Q_2K_2 with respect to the indifference curve of consumer 2. Since the sum of the distances to the three sides of P^sRS is equal to one for all points of this triangle, it is clear

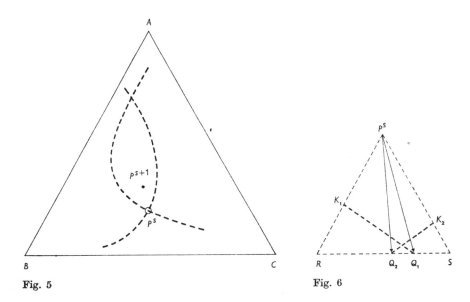

Fig. 5   Fig. 6

Hence the procedure has the pleasant characteristic that both individuals benefit from the revision of the program. In this sense it is equitable. Moreover one intuitively understands that, if revisions are made by infinitesimal steps, it converges to an optimum, a property that may indeed be proved rigorously.[1]

6. Correct Revelation of Preferences?

(i) *General Considerations*

The theoretical literature dealing with public goods has placed great emphasis on the problem of the revelation of preferences. P. A. Samuelson has forcefully argued that there was a basic difference in this respect between private and public commodities.[2] By his demands on the market each individual reveals his needs and wants for private commodities. But he sees no reason for reporting correctly his demand for collective consumption, because the latter will be provided even if he does not ask for it.

The planning approach provides the appropriate theoretical framework for a new discussion of the problem. Indeed, the question is to know how the individual will report during the process of exchange of information if he does

that $\pi_1^s + \pi_2^s$ will exceed 1 precisely when Q_1 will be on the right of Q_2. Then there will exist programs with a larger collective consumption that will be above the two indifference curves passing through P^s. To find such programs it is enough to move upwards in a direction opposite to a vector contained in the angle $Q_2 P^s Q_1$, in particular to the vector $P^s \overline{Q}$ where $\overline{Q}$ is the middle of $Q_1 Q_2$. But the distance from $\overline{Q}$ to the side $P^s R$ is equal to the average of π_2^s and $1 - \pi_2^s$, hence to $-\frac{1}{2}(\pi_2^s - \pi_1^s - 1)$. We find the rule given by the first equation (15).

[1] See J. Drèze and D. de la Vallée Poussin [1] and E. Malinvaud [5].
[2] See for instance P. A. Samuelson [7].

not feel himself bound by the rules justifying this process but wants to influence in his favor the final outcome.

A little reflection about the problem shows, however, that it is quite complex. On the one hand, its treatment will rely on the formalizations introduced by the theory of games and by the study of imperfect competition, both of which have taught us the many difficulties raised by the definition of optimal behavior. On the other hand, serious discussion may arise as to the conceptual framework that is appropriate in the present case: should we stress the case of many consumers or that of a few only? what kind of information should we suppose each individual holds when he decides on the proposal he will report?

Dealing with private consumption we have been used to put emphasis on the situation where many individuals act on the markets. Considering collective consumption we may again take for granted that many persons are interested in it; but we must also remember that those persons do not act directly: they are represented in some way, so that the actual decision process should probably be viewed as taking place among a few representatives. A model with two "individuals" is not obviously inappropriate because these two may represent organized social groups.

The theories of games and imperfect competition have usually studied cases in which each agent has full information on the situation. This may still be an appropriate setting for a discussion on how decisions are actually taken. But it is clearly not suitable when we want to compare the properties of various planning procedures which all assume that information is not a priori centralized.

Facing these many problems I shall not try to do justice to the very complex question of revelation of preferences. But looking at it in a simple minded way will already be instructive and will show that it is not as clearcut as the distinction made by Samuelson would suggest.

(ii) *Lindahl's Procedure and Its Revised Versions*

Let us then consider the Lindahl procedure and study how each individual may report if he does not follow the rules imposed on him but tries to benefit from the procedure as well as he can. It will be convenient to assume first that the individual knows not only his own indifference curves but also those of the other, and moreover that he accepts the hypothesis that the other will exactly follow the rule of the procedure.

Let us refer back to Fig. 3 and focus attention on individual 1. In order to decide on what is best for him, he will determine the "reaction curve" of individual 2, i.e. the demands that 2 will announce for each position of the budget line $A'H$ rotating around H. On this reaction curve 1 will find the point that best suits him; he will then behave in such a way as to achieve this point. The Fig. 7 is supposed to be drawn in the triangle in the same way as was Fig. 3. The reaction curve R_2 of the second consumer joins points like D_2

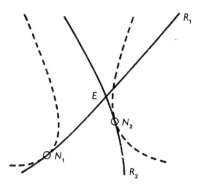

Fig. 7

and D_2' of Fig. 3, points where the indifference curves of 2 are tangent to the budget line. On R_2 the point N_2 gives to 1 the highest utility level. We easily see that it is also the point where, in Lindahl's solution, the first individual will stop accepting an increase in his tax share. If Johansen's planning procedure inspired by Lindahl is considered, then individual 1 will lower his demand for collective consumption in such a way as to report as if his own reaction curve passed through N_2.

Similarly, if the Johansen procedure is followed and if individual 2 assumes that 1 correctly reports along his "reaction curve" R_1, then 2 will aim at achieving point N_1 and therefore lower also his demand of collective consumption. The Samuelson effect on underreporting will be present for both individuals.

The results are not very different if we relax full information as well as the hypothesis that each individual assumes that the other reports correctly. Observing the proposals made by 2, individual 1 can construct a "reaction curve" of 2 and extrapolate it so as to choose on it a preferred point. He will then report in such a way as to aim at this point. If both behave in this way their curves will be below R_1 and R_2 respectively. The outcome of the procedure will be characterized by a collective consumption smaller than the one corresponding to the "equilibrium" point E where R_1 and R_2 intersect.

Increasing the number of individuals will not help, quite the contrary. Taking as given the behavior of other participants individual 1 can still construct in his (z, x_1) plane a path exhibiting the implications for him of the various stages of the procedure. Extrapolating this path he will aim at a preferred point. As the number of participants increases the path will show less and less elasticity to variations in t_1 (and x_1). Hence the amount of underreporting will increase.

The procedure presented in section 4 is not so easily dealt with because the budget line now has two degrees of freedom. Looking, however, at Fig. 4 we understand how individual 1 may distort his preferences if he assumes

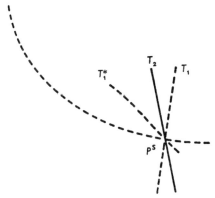

Fig. 8

that 2 will report correctly. He has two ways for obtaining a budget line that will be more favorable to him than PD_1'. He might conceivably report a demand lower than D_2 so that the budget line rotates clockwise; but this would benefit to him only if he could obtain a large quantity of collective consumption, which will not be the case under such a distortion of his preferences. In fact he is most likely to *overstate* his demand in such a way that the budget line rotates around a point located higher than P. In so doing he will impose on individual 2 a "concession" larger than the one he would make under correct reporting.

On the contrary, in the situation of Fig. 4 consumer 2 will find it to be of interest to understate his demand, so that the budget line rotates around a point located below P. Hence distortion of preferences will exaggerate the differences among individuals. This effect will occur whether or not there is full information, whether or not one of the participants is correctly reporting, whether or not there are just two individuals: anyone finding himself wanting more collective consumption than is demanded by the others will have interest to overstate his demand so as to benefit from a more favorable revision of the transfer (even at the cost of a less favorable revision of his tax share). Under such circumstances the convergence of the procedure becomes doubtful.

(iii) *The Procedure with Quantity Indicators*

The situation is radically different for the procedure discussed in section 5 and aiming at the progressive improvement of a feasible program.

The basic reason for the difference is that anyone reporting correctly cannot suffer from the revision of the program, at least if this revision is made by small steps. Indeed, the new program has been defined in such a way that, if π_i^s is the true marginal rate of substitution for individual i, then (x_i^{s+1}, z^{s+1}) is preferred by i to (x_i^s, z^s).

On the contrary, anyone reporting a false π_i^s could loose if other participants pick proposals that are particularly unfavorable to him. Consider for instance Fig. 8 drawn as Fig. 5. Individual 1 indifference curve passing through P^s is exhibited. Assume that individual 1 reports a marginal rate which would imply that his indifference curve is tangent to P^sT_1, thus underreporting his interest for collective consumption. Then individual 2 might report a marginal rate corresponding for his own indifference curve to a tangent P^sT_2 contained between P^sT_1 and the real indifference curve of 1. The program would then be revised by a reduction of collective consumption and in a direction that would reduce the utility enjoyed by 1.

The two preceding propositions imply that, in the framework of the procedure of section 5, reporting correctly defines a minimax strategy for each individual: it is the unique strategy that rules out a decrease of the utility level.[1]

This does not mean that the individual will not try to misrepresent his preferences. He may indeed aim at obtaining a revision that will be particularly favorable to him.

Suppose for instance that individual 1 knows the proposal π_2^s made by 2 and chooses his own proposal π_1^s in order to maximize the utility U_1 that P^{s+1} will imply for him. One can see that π_1^s will then not be quite equal to his true marginal rate of substitution. The two quantities z^{s+1} and x_1^{s+1} are defined according to (14) and (15), the second of which may also be written as:

$$x_1^{s+1} - x_1^s = \frac{b}{2}\left[(\pi_2^s - 1)^2 - (\pi_1^s)^2\right]. \tag{16}$$

The maximum of $U_1(x_1^{s+1},\ z^{s+1})$ will be achieved for the proposal π_1^s that implies:

$$U'_{1x}(x_1^{s+1}, z^{s+1})\frac{dx_1^{s+1}}{d\pi_1^s} + U'_{1z}(x_1^{s+1}, z^{s+1})\frac{dz^{s+1}}{d\pi_1^s} = 0 \tag{17}$$

But (16) and (14) show that:

$$\frac{dx_1^{s+1}}{d\pi_1^s} = -b\pi_1^s \qquad \frac{dz^{s+1}}{d\pi_1^s} = b. \tag{18}$$

Hence the best π_1^s will be given by:

$$\pi_1^s = \frac{U'_{1z}(x_1^{s+1}, z^{s+1})}{U'_{1x}(x_1^{s+1}, z^{s+1})}. \tag{19}$$

It is the marginal rate of substitution evaluated at P^{s+1}, not at P^s. If the indifference curves have the usual curvature π_1^s will be closer to $1 - \pi_2^s$ than

$$\frac{U'_{1z}(x_2^s, z^s)}{U'_{1x}(x_1^s, z^s)}. \tag{20}$$

---

[1] This point was first made by J. Drèze and D. de la Vallée Poussin [1].

On Fig. 8, $P^sT_1^*$ is meant to represent the proposal π_1^s determined according to (19).

If the revision is made by small steps the difference between (19) and (20) will be negligible. But the case of two individuals is particularly favorable in this respect. When we generalize the argument to the case of m individuals, we find that π_1^s will be biassed downward if the sum of the proposals made by all participants exceeds 1, even if the steps are infinitesimal (for $m > 2$), and that the absolute value of the bias will increase with m. However, if all individuals follow the behavior that we are now assuming, the procedure will still converge to an optimum, although more slowly than under correct reporting.[1]

7. Quantity or Price Indicators?

We began our discussion with the solution suggested by E. Lindahl and implying the consideration of price indicators: the tax shares t_i. We modified this solution so that it meets better the requirement for equity. Then we shifted to a procedure using quantity indicators. We just found that the latter is much more robust than the two former ones with respect to the misrepresentation of their preferences by the consumers. The incentives in favor of correct reporting are much less weak.

This theoretical result may be viewed as giving support to the fact that actual debates on collective consumptions are organized on specific projects along lines that seem to be better represented by the last procedure than by the former two.

The case we have been considering is obviously too simple. The fact that there are only two commodities, one public, the other private, is particularly troublesome. However, generalizations exist. In particular the pleasant properties of the last procedure are also found in a model with m consumers, r public goods and $n - r$ private goods if one considers a procedure that uses as indicators quantities for the public goods and prices for the private goods.[2]

References

1. Drèze, J. & de la Vallée Poussin, D.: *A tatonnement process for guiding and financing an efficient production of public goods*. CORE *Discussion Papers, no.* 6922, July 1969.
2. Johansen, L.: *Public Economics*, in particular section 6.2.2. North-Holland,

Amsterdam, 1965. For a very enlightening discussion of Lindahl's solution, see also "Some Notes on the Lindahl Theory of Determination of Public Expenditures", *International Economic Review*, September 1963.
3. Lindahl, E.: "Just taxation—A positive

---

[1] This property was shown to me by C. von Weiszäcker in a personal communication. It implies in particular that, if P^s is optimal, then each individual will prefer to report correctly when all take as given the proposals made by others. This conclusion was also derived directly by J. Drèze and D. de la Vallee Poussin [1].

[2] See E. Malinvaud [6].

solution", first published in German in 1919, reprinted in *Classics in the Theory of Public Finance* (ed. R. A. Musgrave & A. T. Peacock). MacMillan, London, 1967.

4. Malinvaud, E.: Decentralized procedures for planning. In *Activity Analysis in the Theory of Growth and Planning* (ed. E. Malinvaud & M. O. L. Bacharach) Macmillan, London, 1967.

5. Malinvaud, E.: "Procédures pour la détermination d'un programme de consommation collective" (mimeographed, INSEE, 1969) submitted to the *European Economic Review*.

6. Malinvaud, E.: "The theory of planning for individual and collective consumption" (mimeographed, INSEE, January 1970).

7. Samuelson, P. A.: Pure theory of public expenditure and taxation. In *Public Economics. An Analysis of Public Production and Consumption and their Relations to the Private Sectors* (ed. J. Margolis & H. Guitton). Macmillan, London, 1969.

AN APPROACH TO THE PROBLEM OF ESTIMATING DEMAND FOR PUBLIC GOODS

*Peter Bohm**

University of Stockholm, Stockholm, Sweden

The purpose of this paper is to advance a procedure for estimating the demand for public goods (such as environmental quality) in the probably normal case where the authorities have no idea about the aggregate maximum willingness to pay in relation to costs nor about the willingness to pay in different consumer groups which are possible to distinguish from the taxation point of view. It is argued that the procedure advanced here will avoid creating incentives to misrepresent individual preferences. Moreover, it conforms to the real-world situation where an *optimal* payment or taxation distribution cannot be decided upon *prior to* estimation of demand.

Our approach is presented in Section II. In the final section, it is placed into the context of the pure theory of public goods as formulated in a recent article by McGuire & Aaron [3].

I. Introduction

During the recent development of the theory of public goods it has been repeatedly said that we cannot expect consumers to reveal their true preferences for such goods. The well-known reason is that the rational individual knows that the effect he might have on aggregate demand, and thus on the amount produced of the public good, is negligible whereas he can achieve a nonnegligible pay-off in terms of a reduced payment for these goods by stating a willingness to pay that falls short of his true valuations.[1]

On the other hand, the actual practice in this field, if any, has been quite to the contrary. By the formation of public opinion within or by means of political parties, by the creation of pressure groups and in the form of polls taken, potential consumers of a proposed output of a public good have stated preferences which can only be expected to *over*estimate their true valuations. The simple reason is, of course, that the consequences as to their payments (e.g. a tax increase) have been left out of the process.

If the maximum willingness to pay with respect to a specific public good

---

* The author is indebted to Professor Edmond Malinvaud, Mr Alf Carling, Mr Leif Magnusson and Mr Lewis Taylor for many helpful comments on an earlier version of this paper.
[1] Cf. e.g. Samuelson [5], Musgrave [4], p. 80.

proposal were estimated by means of some version of each of the two approaches just mentioned, i. e. one in which charges are levied according to the (marginal) willingness to pay, and another in which the good is understood to be made available free of charge to the individual, we should expect the true maximum willingness to pay to lie somewhere in between the two estimates received. Such a double estimate, when feasible, would in itself be a step forward from having only an upper or a lower bound to the demand. But the fundamental question is whether there exists one single approach which gives a reasonably accurate estimate of actual demand.

The approach to estimation of the demand for public goods which will be discussed in this paper stems from a straightforward observation of the two approaches mentioned above: Both put the individual in a situation where the choice of strategy is very simple. Then, one possible way to achieve a solution lying between the two extremes would be to make the formation of strategies more difficult, preferably so difficult that the individual would tell himself something like: "Since I cannot find a way to beat the system, I had just as well tell the truth. Then, at least I can't lose."

Even when our approach could not be expected to achieve this ideal response from every individual, valuable information might still be obtained. This would be the case if, for example, there were no *systematic* bias in the strategic behavior of the individuals, that is, if it could be shown that people would respond in such a way that understatements and overstatements were stochastically distributed. Under these circumstances, the aggregate stated willingness to pay would be a good estimate of the sum of the consumers' true valuations.

II. A "Counter-Strategic" Approach

II.1 *A simplified exemple*

Assume that a proposal to proceduce a certain amount of a given public good has been put forth — for that matter, it might even concern a private good with decreasing average costs and with marginal cost pricing. As a specific example, let us consider a plan to build a waste treatment plant according to a given technical solution known to make it possible to go swimming and fishing in a particular lake, now unsuitable for these purposes. Let those who can possibly benefit from its construction[1] be confronted with a question about their maximum willingness to contribute to the construction of this plant. (For example, the question might be posed on the form for the income-tax return.) The potential beneficiaries are informed that, if the total willingness to pay exceeds the given construction costs (maintenance costs assumed to be zero), the plant will actually be constructed. Moreover, they *may* be called upon to pay exactly the amount they have stated. They are, however, also informed that

---

[1] For simplicity, let us assume that no one will be negatively affected by its construction.

the actual payment may not be equal to this amount, that, for example, it may be proportional to the amount stated. As additional aspects of fiscal administration, of distribution etc. will influence the choice of method of payment, it is also possible that everybody will be called upon to pay the same "price" (at most $ X). Furthermore, it is possible that the costs of the plant will be partly or completely financed by the federal government and thus only partially or not at all by the people in the region concerned. Other alternatives may be explicitly mentioned as well. Anyway, it is made clear that, at the moment of the "referendum", the actual distribution of payment is unknown.

II.2 *The credibility gap*

Even this simplified example involves several problems. Perhaps the most serious one is that of a "credibility gap". The consumers may believe from the information received that the "game" is a fraud and that the administration, the "local government", actually knows which payment alternative will be applied if the treatment plant is built, but will not say so for some reason. Fortunately—at least from the point of view of our present interest—this gap may be closed by making a few points of clarification. To begin with, it may be pointed out (1) that, historically, relevant items of local government activity have in fact been financed in a number of different ways. More important, (2) certain factors decisive for the choice of payment alternative will only be made known from the *results* of the referendum, factors such as how the willingness to contribute is divided among subregions (each of which may have a local government), among age groups, income groups, family sizes etc. (all relevant for the welfare-distribution aspect of the problem) and among different types of taxpayers (which will influence the choice of payment procedure). (3) It can be said, if true, that the consequences of alternative payment procedures in the light of the results from the referendum will be studied by a group of non-political experts. Thus, the final set of alternatives, from which the government will choose one, will be made public after the referendum. Finally, (4) the mere fact that the local government poses this question to the people indicates that the government truly does not know the preferences of the consumers (otherwise, for obvious reasons, the referendum would not be arranged). This contributes to making it *likely* that the final payment procedure is in fact unknown.

If this uncertainty were not to obtain, if, for example, there were plans to select one particular alternative or if quite different and approximately known probabilities could be attached to the alternatives, then the referendum from our point of view would either be called off or formulated in another way. Our approach, thus, has nothing to do with bluffing, not necessarily for moral reasons, but for the practical reason that otherwise this referendum or future ones would provide less accurate information.

II.3 *Strategic elements in the formation of rules of the referendum*

If some ranking or other specification of the probabilities of the payment alternatives can be made, there may be another way to formulate the proposition. In the trivial case where some alternatives can simply be ruled out, we are back at our first version, only with fewer alternatives. If, on the other hand, the probablities are positive and it is possible to rank them with respect to informed guesses about the behavior of the electorate, the attitude of the federal government etc., then a second version of the referendum might include a presentation of these arguments, so that the voters will be informed about the beliefs of certain persons speaking in the name of the government.

This second version brings out a general problem, which it does not seem possible to settle in a completely satisfactory manner. Just because of the uncertainty of the actual payment alternative, which is the essential element of our approach, we cannot disregard the fact that individual members of the government would be well protected, if they chose to express beliefs not actually held by them. As problems of this nature may dominate in the second version, we shall exclude it in this paper. However, it is quite possible that, in a world with extensive uncertainty about preferences for certain public goods, it is better to have access to the results of an imperfect referendum than to be completely without information of this kind.

Although exclusion of the second version reduces the problem just brought up, it does not eliminate it. In the first version, individual members of the government who, say, want to have the waste treatment plant built and later will do whatever they can to have a certain payment alternative selected may gain by saying that the choice of payment alternative is *completely* unknown. But as long as the positions of all or most members are not compiled, it would be correct to assume that the position of the majority is in fact uncertain. This assumption, which in fact seems realistic in the situation under consideration, becomes even less restrictive if the (local) authorities who arrange this referendum have only limited power over the actual payment alternative to be chosen. This is the case when the "federal" government takes the position that it is willing to discuss contributions to local public projects but that it, as a matter of principle (if for no other reason), cannot say what its position will be before the results of the referendum are known. Thus, we end up in a situation in which a large number of individual members of more than one political body will determine the final choice of payment alternative with an additional, unknown element of influence from experts in relevant government committees and government agencies. This, in short, is the situation we shall have in mind.

II.4 *Incentives to distorted responses*

To sum up the argument so far, we have a situation where the authorities do not know whether or not the costs of a specific public good proposal exceed

the aggregate willingness to pay for this output and where every precaution-
ary measure has been taken to inform people that there is truly no way of
knowing which of a given set of payment alternatives will hold. In this situation,
it can be assumed that the individual consumer regards the outcome of the
referendum and the payment structure later to be adopted, when relevant, as
completely uncertain.

Our argument, then, is simply that the incentives for the individual consumer to distort statements of his maximum willingness to pay in this case
are far weaker than when any one payment alternative is known to hold. In
particular, if the possible payment structures include alternatives such as
(a) paying the amount corresponding to the stated maximum willingness to
pay, (b) no payment at all (federal funds used for a local project), etc., it is
assumed that the consumer knows that he would benefit from understating
his willingness to pay if case (a) were *known* to hold, and overstating it if
case (b) were *known* to hold. But, when these alternatives along with others
are all possible and have unknown probabilities, the incentives to distort the
statements are no longer unambiguous.

The possibility now arises that the incentives to, say, understatements are
stronger than the incentives in the opposite direction. For instance, when the
number of beneficiaries is very large, uncertainty cannot be expected to eliminate completely the "free-rider" incentive. That is to say, the individual consumer knows that, if he states his true willingness to pay, he *may* end up in a situation in which he pays more than "necessary" for the given public good output.
But neither can he neglect the fact that he may have *some* influence on the
final decision whether to produce the good or not *and* that he may have to
pay very little, perhaps nothing, for this influence. It is hard to predict his
behavior under these circumstances. In particular, the individual who in a
given situation *knows* that he would gain from understating his preferences and
takes advantage of that knowledge, may not wish to do so in this situation of
uncertainty. For example, as has been argued, the individual may have an
interest in a "good" society as well as in his own direct personal well-being,[1]
and the former interest may gain in weight relative to the latter, when the
benefits of "cheating" are reduced or made less certain.

Once it is clear that there is no open-and-shut case for the individual when
considering under- or overstatement of his preferences, the choice of strategy
may well seem so complicated to him that he prefers to state his true maximum
willingness to pay. Only an actual test of alternative approaches to estimating
the demand for public goods could reveal the true state of affairs in this respect.[2]
But one important argument in favor of the hypothesis that people will abstain
from the complicated calculations of optimal strategy inherent in our approach

---

[1] Cf. Musgrave [4], p. 87–88, and Marglin [2].
[2] The results of an experiment along these lines will be presented in a forthcoming paper
[1].

is that most people simply won't find such calculations worth while considering the small individual sums usually involved.[1]

Nevertheless, if it is shown that a net incentive for, say, understating preferences will prevail, this tendency may be corrected in more that one way. One might specify additional possible payment alternatives which, if known to hold with certainty, would provide incentives for overstatements, for example, certain small payments equal for all. Or the referendum might be reduced to a "hearing" of a random *sample* of potential beneficiaries, in which each participant would feel a stronger influence on the aggregate maximum willingness to pay.

This approach using a random sample may be appropriate for other reasons as well. A referendum may turn out to be impossible in practice on institutional grounds or because of high costs. Even if no such obstacles existed, polling all potential consumers may still turn out to be undesirable. Say that it is judged to be impossible, for whatever reason, to collect a specific sum from each individual, but that it is feasible to do so from different *groups* of individuals, e.g. different income brackets and/or different categories of consumers. Then, the appropriate approach would be to use a representative sample with a sufficient number of people in each relevant category. On the basis of the responses from this sample a particular payment procedure is chosen for the population as a whole (a fee equal for all, different fees for different categories, a proportional or progressive increase in income taxes, etc. or combinations of such schemes). The persons in the sample may, for the sake of proper incentives, be given special treatment, say, by never being forced to pay more than their stated maximum willingness to pay.

Needless to say, this application of our approach to a sample instead of to the entire relevant population would be optimal in those cases where it were considered possible to divide the citizens into groups of (approximately) identical preferences, i.e. when preferences could be assumed to be the same within different groups with respect to age, income, type of occupation, location etc. When we now relate our approach to the pure theory of public goods, we shall regard the consumers as representatives for different groups of persons with identical preferences, with groups of one consumer only as a possible but special case.

III. The Pure Theory of Public Goods

Our starting points can be summarized as follows. (1) We argued in the preceding section that, in certain important situations, i.e. where data about consumer preferences are most badly needed by the policy makers, there exists a practicable method to estimate the demand for a public good in which

---

[1] This is also a reason for rejecting any expected utility approach to analyse behavior in this context. Another reason is that people are unlikely to react as if they could attach probabilities to each of the different payment alternatives.

incentives to misrepresent preferences are weaker, if at all present, and strategy calculations much more "expensive", if at all relevant or meaningful under the given state of uncertainty, than in other approaches discussed in the literature on public goods. (2) We have limited the discussion to the case of a specific public good proposal, which may refer to an indivisible or a divisible good. In the case of a divisible good, it is assumed that a realistic and practicable approach to estimation of the maximum willingness to pay calls for a strictly limited number of alternative output levels, perhaps only one. (3) We have suggested that, given a set of specific payment alternatives, the final choice will be influenced by aspects of welfare distribution, policy costs (to be defined) and policy constraints.

The purpose of the present section, then, is to analyse optimal policy behavior subject to certain general distribution considerations, policy constraints and policy costs, when there exists a proposal to produce a certain output of a public good and when there exists an estimate of individual maximum willingness to pay for the given output. For this purpose, we shall use a model presented by McGuire & Aaron [3].

III.1 *The McGuire-Aaron Model*

McGuire & Aaron have shown how various constraints on economic policy in a two-person economy with one public good (x) and one private good (y) will affect the optimal output of the public good and whether or not the conditions for Pareto-optimal allocation are fulfilled at this output. Starting from a situation of $x=0$, where the national income (y) is divided between the two persons a and b, such that $y_a^I + y_b^I = y$, the optimum level of x is sought. In addition to x, there are two policy parameters potentially available: (1) Income (or, preferably, lump-sum) taxes and transfers which are viewed as shifts in y_a^I, a's income after general taxes and transfers. (2) A special fee to be charged for the public good, which may be seen as a "price" paid per unit of x, denoted by p_i ($i=a,b$). For a given set of y_i^I, x and p_i and for individual preferences described by $U_i(y_i^c, x)$, where y_i^c is i's consumption of the private good ($=$ net income after *all* taxes, transfers and fees), the situation of the individual consumer is defined by

$$U_i(y_i^c, x) \qquad \text{subject to } p_i x + y_i^c = y_i^I.$$

McGuire & Aaron assume that the marginal cost of x is constant $=p$, that fixed costs are zero and that there is a balanced "allocation" budget such that $\Sigma p_i = p$. Moreover, there is a social welfare function, $W = U_a + \gamma U_b$ where γ is a positive constant. The general problem is then to maximize

$$W = U_a(y_a^I - p_a x, x) + \gamma U_b[(y - y_a^I) - (p - p_a) x, x],$$

which is carried out for the following three cases:

(a) p_a, y_a^I are given ($y_a^I = \bar{y}_a^I$); x is variable. The first-order optimum condition in this case is simply:

$$\frac{\partial W}{\partial x} = -U_{a1}p_a + U_{a2} + \gamma(U_{b2} - (p - p_a)U_{b1}) = 0 \tag{1}$$

The optimum feasible solution here implies that the condition for Pareto optimum is not fulfilled, i.e., the sum of the marginal rates of substitution ($\sum \text{MRS}_i$) $\neq p$, except by chance.

(b) y_a^I or p_a is given; x and either p_a or y_a^I are variable. The first-order optimum conditions are now (1) and (2) *or* (1) and (3):

$$\frac{\partial W}{\partial y_a^I} = U_{a1} - \gamma U_{b1} = 0 \tag{2}$$

$$\frac{\partial W}{\partial p_a} = -U_{a1}x + \gamma U_{b1}x = 0 \tag{3}$$

Equation (3) is relevant only for $x > 0$ and thus identical to (2) provided the constraint $0 \leqslant p_a \leqslant p$ is ineffective. In this case (1) and (2), or (1) and (3), imply that $\Sigma \text{MRS}_i = p$, i.e. a first-best optimum.

(c) p_a, y_a^I and x are all variable. The optimum conditions are now (1), (2) and (3). Clearly, the same result as in (b) is obtained and one of the policy parameters p_a and y_a^I is redundant.

III.2 *A given supply of public goods*

In all three cases above x was variable, whereas our approach deals with a specific public good proposal, $x = \bar{x}$. We shall therefore analyse situations in which condition (1) above is no longer relevant and in which x equals either 0 or $\bar{x}$.

Let us start by postulating that each individual shall pay at most what amounts to his maximum willingness to pay. Hence, denoting his consumption of the private good by y_i^c when $x = \bar{x}$, we require that $y_i^c \geqslant \tilde{y}_i^c$, where $\tilde{y}_i^c$ is defined by $U_i(\tilde{y}_i^c, \bar{x}) = U_i(\tilde{y}_i^I, 0)$, $\tilde{y}_i^I$ being his original private consumption. This condition means, of course, that nobody should be made worse off by the introduction of the public good. In addition, for $\bar{x}$ to be preferred to $x = 0$ we must have $y_i^c \geqslant \tilde{y}_i^c$ for at least one individual. This means, that for $\bar{x}$ to be preferred we require $y_i^c \geqslant \tilde{y}_i^c$ and

$$\sum y_i^c > \sum \tilde{y}_i^c \quad \text{or} \quad \sum(y_i^I - p_i\bar{x}) > \sum \tilde{y}_i^c \quad \text{or} \quad y - \sum \tilde{y}_i^c > p\bar{x}, \tag{4}$$

where y_i^I is taken to be the total income (as defined) of the i'th individual at $x = \bar{x}$. This condition (4), derived from our postulate of $y_i^c \geqslant \tilde{y}_i^c$ with strict inequality for at least one individual, is the basis for our criterion in Section II, that $\bar{x}$ should be chosen only if the total willingness to pay exceeds the cost of the public good.

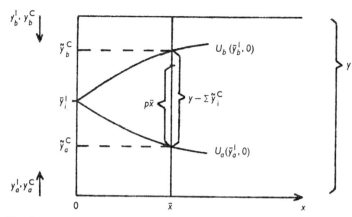

Fig. 1

It is obvious that our approach here is relevant for such changes in the output of public goods where there are policy restrictions which imply—intentionally or unintentionally—that nobody is to be made worse off.[1] Another case is that of *no* constraints on changes of the income distribution at $x=0$ such that condition (2) above is fulfilled at $\bar{y}_i^I$ (see Fig. 1).[2] Now, as an optimal income distribution can also be arrived at when $x=\bar{x}$, we must have for $W_{x=\bar{x}} > W_{x=0}$ that $y_i^c \geqslant \tilde{y}_i^c$ for all i and $y_i^c > \tilde{y}_i^c$ for at least one i, and vice versa; hence, condition (4) is clearly seen to be relevant in this case as well.[3] In fact, where it is possible to increase the welfare of one individual by $y_i^c - \tilde{y}_i^c$ without decreasing the welfare of any other individual, all will share this increase in accordance with the preferences revealed by the W-function.

Our approach is thus relevant for two formulations of the allocation problem: One is to maximize W, where the income distribution could be given any desired value. The second formulation is to maximize W, when there are restrictions on the choice of income distribution and perhaps also on the special fees p_i, above that of $p \geqslant p_i \geqslant 0$, to imply whatever we like but at least that nobody is to be made worse off in the process of introducing the public

---

[1] It seems quite reasonable to assume that there are such intentional constraints on economic policy stemming from the fact that certain groups, whose income the *government* wishes to decrease in favor of other groups, are backed by, say, a majority in the *parliament*.

[2] In this open-end box diagram, used by McGuire & Aaron, the vertical axis indicates the national income (y) which is divided between a and b, and the horizontal axis the amount of the public good. The indifference curves U_a and U_b (drawn for the case of a divisible good) refer to the original position and are taken to reflect the optimal welfare distribution at $x=0$.

[3] Introducing condition (4) in Fig. 1, we have that the distance between the indifference curves at x, i.e. $y - \Sigma \tilde{y}_i^c$, should exceed $p\bar{x}$, the amount of y required for the production of $\bar{x}$. The end points of the segment $p\bar{x}$ in the Figure can then be interpreted as indicating a particular welfare distribution where the case illustrated in the Figure implies that $y_a^c = \tilde{y}_a^c$ and $y_b^c > \tilde{y}_b^c$.

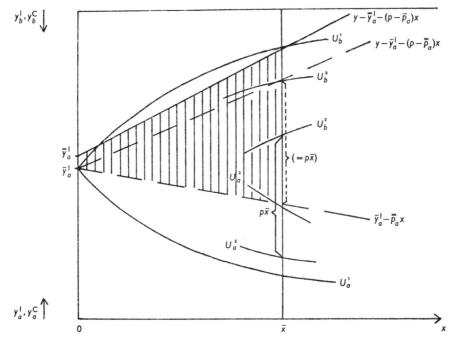

$y_b^1, y_b^C \downarrow$

$y - \bar{y}_a^1 - (p - \bar{p}_a)x$

U_b^1

$y - \bar{y}_a^1 - (p - \bar{p}_a)x$

U_b^3

U_b^2

$\bar{\bar{y}}_a^1$

$\bar{y}_a^1$

$\} (= p\bar{x})$

$p\bar{x} \{$

U_a^3

$\bar{y}_a^1 - \bar{p}_a x$

U_a^2

U_a^1

$y_a^1, y_a^C \uparrow$

0         $\bar{x}$ x

Fig. 2

good. The optimum conditions of the first problem are (2) and (4) from above. The same optimum conditions hold for the second problem if the constraints do not turn out to be effective. Otherwise we must use the parameters to one extreme, say to $p = \bar{p}_a$ and $y_a^1 = \bar{y}_a^1$. An example of a situation where constraints are effective is shown in Fig. 2 where the first-best optimum position is represented by (U_a^2, U_b^2) and where y_a^1 is restricted to the interval $[\bar{y}_a^1, \bar{\bar{y}}_a^1]$, and p_a to the interval $[\bar{p}_a, \bar{\bar{p}}_a]$, giving the best feasible situation as U_a^3 and U_b^3.

In both versions the solutions are attained in the following way: The initial situation given, the maximum willingness to pay is determined for all i. If the aggregate willingness to pay is large enough to pass the test of (4), then $\bar{x}$ should be produced. What remains now is the pure problem of welfare distribution, in the second version bounded by the policy constraints on y_a^1 and p_a. In terms of Figs. 1 and 2, this means that the segment $p\bar{x}$ is shifted between the original indifference curves at $x = \bar{x}$ so that the optimum feasible solution is attained.

Looking at the distribution problem from a practical viewpoint, it is the estimate of the willingness to pay of each individual $(\bar{y}_i^1 - \tilde{y}_i^c)$ that provides the basis for a policy decision. In this way the decision makers at least know how people rank the introduction of the public good in terms of their original incomes.

III.3 *Policy costs*

The fact that the choice of payment alternative (y_i^I and/or p_i) is based on the statements of individual willingness to pay ($\bar{y}_i^I - \tilde{y}_i^c$) is one reason for our assumption in Section II that the payment alternative to hold was unknown at the time of the inquiry into the consumers' willingness to pay. A second reason will appear when we now introduce the costs of policy making (or collection costs), i.e. costs which may vary with the different policy parameters and even with the values of each parameter.

The solution to the earlier problem with these costs neglected has given a certain change in y_i^I and a certain p_i. If a pair of parameter values, substantially lower in terms of policy costs, is found to exist and judged to involve an acceptable departure from the (so far) optimal solution, it will be chosen. It is, of course, not possible to make this choice beforehand as it is based not only on *a priori* knowledge, i.e. on knowledge of the structure of administrative costs, but also on the size and distribution of the aggregate willingness to pay. Again, it is only after these additional data are known that the (overall) optimal payment system can be selected. (We should add here that in reality there are, of course, forms of taxes and charges other than the two observed in the simple model above that it may be relevant to consider from the point of view of policy costs.)

It is only when the costs of various payment alternatives differ substantially and when W is maximized with respect to these costs that the possibility of having $y_i^c < \tilde{y}_i^c$ for some i enters our model. (For example, this could be the result if everybody were called upon to pay the same fee, $p_a = p_b$ and y_i^I unchanged at $\tilde{y}_i^I$.) This possibility was excluded in the pure distribution part of the problem by the explicit requirement, $y_i^c \geqslant \tilde{y}_i^c$. The reasons for such a requirement at that stage of the analysis must now be spelled out. First, many voters would be likely to suspect how a redistribution in which some consumers could be made worse off would affect them. This means that some voters, say, high-income earners, would feel that they run a large risk of being forced to pay sums exceeding their maximum willingness to pay, whereas others may know that they belong to a group to which the government wishes to make income transfers and that they therefore probably will not have to pay very much. The first group of voters may have a tendency to understate their preferences, to do what they can to stop the production of the public good, and vice versa. These distortions not only jeopardize the information value of the individual responses but also the information value of the aggregate willingness to pay as it will be difficult to determine the resulting net bias. In contrast, *policy-cost* considerations leading to $y_i^c < \tilde{y}_i^c$ for some i would hardly have effects of this kind as it seems far-fetched to assume that the individual consumer would have an idea about the direction in which such considerations would affect him.

Second, if we were to admit intentional welfare redistributions where

$y_i^c < \tilde{y}_i^c$ for some i, we would not always be able to formulate a meaningful criterion for the decision as to whether $\bar{x}$ is to be produced. With constraints on economic policy, criterion (4) would then be quite arbitrary, as it might be preferred to have $x = \bar{x}$ even if (4) were not fulfilled, and vice versa. Moreover, the relevant criterion could not then be determined before the (true) preferences were known. It must be judged as unsatisfactory from the point of view of information quality to ask people to estimate their willingness to pay without telling them under what circumstances the good will be produced. And in this respect, (4) is a very simple criterion and one which the voters can be expected to consider as natural or reasonable.

Then, having introduced the policy-cost aspect, we can describe the choice of optimal policy as follows: $\bar{x}$ is to be produced if the aggregate willingness to pay exceeds production costs and optimal policy costs. Then, given the statements of all consumers, W is maximized for $x = \bar{x}$ with respect to y_i^I and p_i, subject to $y_i^c \geqslant \tilde{y}_i^c$. Introduce the functions describing the policy costs of the parameters. Investigate the possibilities for *decreasing* these costs by departing from the above-mentioned pair of y_i^I and p_i. If there are such economies to be made which exceed the imputed loss in terms of a (possibly) less desirable welfare distribution, choose the best feasible alternative. Now, the optimum policy costs and hence the total costs of x can be calculated. If (4) is fulfilled at these costs, $\bar{x}$ should be produced.

The reason for introducing the policy costs here is most certainly not that they can frequently be expected to have a significant effect total costs and hence on the choice whether to proceduce the good or not. The fact that the costs for all relevant payment systems may be small in relation to the production costs does not, however, mean that they are unimportant in the context of choosing an optimal payment system.

To conclude, both distribution and policy cost considerations make the optimal system of individual payments for a public good unknown at the time of the inquiry into the willingness to pay for this good. Hence, the procedure suggested in Section II both applies to a realistic policy situation and avoids simple strategies for misrepresentation of preferences.

References

1. Bohm, P.: Estimating Demand for Public Goods: An Experiment. (To appear.)
2. Marglin, S. A.: The social rate of discount and the optimal rate of investment. *Quarterly Journal of Economics*, 1963.
3. McGuire, M. & Aaron, H.: Efficiency and equity in the optimal supply of a public good. *Review of Economics and Statistics*, 1969.
4. Musgrave, R.: *The Theory of Public Finance*, New York, 1959.
5. Samuelson, P.: The Pure Theory of Public Expenditure. *Review of Economics and Statistics*, 1954.
6. Samuelson, P.: Pure theory of public expenditure and taxation. In *Public Economics* (ed. Margolis & Guitton), London, 1969.

A METHOD OF ESTIMATING SOCIAL BENEFITS FROM POLLUTION CONTROL

Karl-Göran Mäler*

The Stockholm School of Economics, Stockholm, Sweden

I. Introduction

In an article in *Water Resources Research*, 1966 [3], Joe B. Stevens tried to estimate direct recreational benefits from water pollution control by using market demand curves for a sport fishery. The quality of the fishery was represented by the angling success per unit of effort. Water pollution would cause a deterioration in the quality, i.e. would decrease angling success. By estimating a demand function for the sport fishery, both as a function of the price of using the fishery and as a function of the quality variable, Stevens thought he could calculate the recreational benefits or the willingness to pay for maintaining constant quality, from various areas under the demand curves.

Stevens' idea, although a very sound one, was not developed in a rigorous way and his conclusions were therefore vague. The aim of this article is to develop a theory which can lend support to calculations such as those presented by Stevens.

The ideas in this article will first be presented intuitively in a non-rigorous way. Then Section III includes a brief review of elements from demand analysis and a statement of the marginal conditions for Pareto-optimality in an economy with public goods. A theoretical framework is developed in Section IV which enables derivation of the willingness to pay for public goods in certain cases on the basis of information on demand functions for private goods.

II. Intuitive Presentation of the Main Idea

It is natural to assume that if it is known that a public good is complementary to a private good, then it should be possible to calculate the demand for the public good if the demand for the private good is known. And if a public good is a perfect substitute for a private one, the consumers' preferences for the public good can be derived from the revealed preferences on the market.

A systematic study of the *a priori* conditions with regard to the preferences for a public good, obtained from complementariness and substitutability is

* I am very grateful to Professors P. Bohm, Clark Reynolds and Robert Solow for valuable comments and suggestions. This project was supported by the Ford Foundation and Stiftelsen Riksbankens Jubileumsfond.

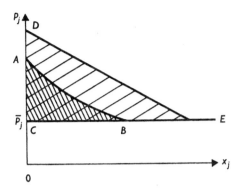

Fig. 1

required. We will not undertake this kind of study here, but instead discuss a single condition which seems realistic in some cases.

Consider a private good x_j, which can be produced in different qualities Y_k, but in only one quality at a time. The use of a fishery, for example, can be regarded as a private good. The quality of the fishery measured in terms of the catch of fish per unit of fishing time or the oxygen dissolved is a public good in the sense that all consumers using the fishery meet with the same quality. Those who do not use the fishery will generally be indifferent to quality changes. A single consumer will be indifferent to quality changes if the price of the corresponding private good is high enough to prevent him from using the fishery. This idea, simple as it is, provides an additional condition which can used for calculating the demand price for quality changes. (Note that if the private good can be supplied in different qualities at the same time, the quality ceases to be a public good. In this case we do not have one private good but many, each characterized by a certain quality.)

The condition can be stated mathematically in the following way: There exists a private good x_j and a public good Y_k such that

$$\frac{\partial u^h(x_1, \ldots, x_{j-1}, 0, x_{j+1}, \ldots, x_n, Y_1, \ldots, Y_m)}{\partial Y_k} = 0 \tag{1}$$

where u^h is the utility function of consumer h.

Assumption (1) implies that the demand price for Y_k can be calculated from the demand function for x_j. This can be observed intuitively as follows.

Consider the compensated demand curve AB for good j in Fig. 1. At the price $p_j = \bar{p}_j$, the consumer demands $\bar{x}_j$ and the consumer surplus is the cross-shaded area ABC. The consumer is thus willing to pay ABC in order to prevent a fall in the supply from $\bar{x}_j$ to zero.

Now consider a change in Y_k. This will cause the compensated demand curve to shift to DE. The new consumer surplus becomes the area CDE

(provided the price does not change). How much is the consumer willing to pay for the change in Y_k, that is how much is he willing to pay for the induced movement from B to E?

This movement can be divided into three steps:

(a) a change in p_j from C to D. In order not to put the consumer in a worse position, he has to be compensated by ABC.

(b) a change in Y_k. If we apply assumption (1), this change will not cause any need for compensation.

(c) a change in p_j from D to E. In order not to put the consumer in a beter position, he has to pay the amount CDE.

The net result is $CDE - ABC$ or the area $BADE$. The amount the consumer is willing to pay to obtain the change in Y_k is thus $BADE$.

Note that this calculation is impossible without assumption (1). If (1) is not applicable, the appropriate transfer in step (b) cannot be estimated. If the consumer is willing to offer something in order to improve the quality of x_j even if he does not consume x_j, then nothing can be said about his willingness to pay for a change on quality on the basis of his demand curve for x_j.

III. Summary of Results from Demand Analysis

This section contains a brief review of some elementary parts of demand analysis which are relevant to this study.

We assume that there are H consumers, each equipped with a utility function

$$u^h(x^h, Y) \quad h = 1, ..., H$$

where x^h is the vector of net demand for private goods (there are n private goods in the economy) and Y is the vector of public goods supplied (there are m public goods and each public good is characterized by the condition that an increase in the supply of the public good for one person means an identical increase in the supply to all other persons). In this context environmental quality is interpreted as a public good because the quality of the water in a stream is the same for everyone.

We assume that $u^h(x^h, Y)$ is twice continuously differentiable and that u^h is quasi concave. We denote the partial derivatives by subscripts:

$$\frac{\partial u^h}{\partial x_i} = u_i^h$$

$$\frac{\partial u^h}{\partial Y_k} = u_k^h.$$

Given the lump sum income I^h and the price vector $p \in R^n$, the budget set for the h:th consumer is defined by

$$M^h = \{x^h \mid p^T x^h \leqslant I^h, x^h \geqslant 0\}.^1$$

We can now study the two "dual" problems

max $u^h(x^h, Y)$

s.t. $x^h \in M^h$

and

min $m^h = p^T x^h$

s.t. $u^h(x^h, Y) \geqslant \bar{u}^h$.

The first order conditions for the problems are

$$u_i^h = \lambda^h p_i \quad i = 1, ..., n \tag{2}$$

and

$$\alpha^h u_i^h = p_i \quad i = 1, ..., n \tag{3}$$

where λ^h and α^h, respectively, are Lagrange multipliers for the two problems. Solving the first order conditions for the first problem yields:

$$x^h = x^h(p, I^h, Y)$$

$$u^h = u^h(x^h(p, I^h, Y), Y) = v^h(p, I^h, Y)$$

and for the second problem

$$x^h = x^{h+}(p, \bar{u}^h, Y)$$

$$m^h = p^T x^{h+}(p, \bar{u}^h, Y) = m^h(p, \bar{u}^h, Y).$$

x^h is the usual Marshallian demand function and v^h is called the indirect utility function. x^h is the Hicksian compensated demand function and m^h is called the expenditure function.

We need the following results (for proofs see e.g. Karlin, Ch. 8 [1])

(a) m^h is a concave function in p

(b) $x^h(p, m^h(p, \bar{u}^h, Y), Y) = x^{h+}(p, \bar{u}^h, Y)$, that is, if income is varied so that the consumer is always on the same indifference curve, then the compensated demand functions are obtained from the Marshallian demand functions.

---

[1] A vector x is interpreted as a column vector and transposition of a vector to a row vector will be denoted by the symbol $\top$. We use the following conventions for vector inequalities:
$x \geqslant 0$ if $x_i \geqslant 0$ for all components x_i
$x \geqslant 0$ if $x \geqslant 0$ and $x \neq 0$
$x > 0$ if $x_i > 0$ for all components x_i.

(c) the Slutsky equations:

$$\frac{\partial^2 m^h}{\partial p_i \partial p_j} - \frac{\partial x_i^h}{\partial I^h}\frac{\partial m^h}{\partial p_j} - \frac{\partial x_i^h}{\partial p_j} = 0, \quad i,j = 1,...,n. \tag{4}$$

The Slutsky equations give the expenditure function as a solution to a system of (not independent) differential equations. The boundary conditions are

$$m^h(\bar{p}, \bar{u}^h, \overline{Y}) = I^h$$

$$\frac{\partial m^h(\bar{p}, \bar{u}^h, \overline{Y})}{\partial p_i} = x_i^h(\bar{p}, I^h, \overline{Y}) \quad i = 1,...,n$$

where $\bar{u}^h = v^h(\bar{p}, I^h, \overline{Y})$ and $\bar{p}, \overline{Y}$ are the prices and the supply of public goods in the initial situation.

If the Marshallian demand function $x^h(p, I^h, Y)$ are known, we can solve the Slutsky equations and determine m as a function of p. But without further assumptions m as a function of Y cannot be determined.

Let us now turn to the problem of aggregation of expenditure functions. The demand functions for consumer h are

$$x^h(p, I^h, Y).$$

The aggregate demand functions are

$$X(p, I^1,..., I^H, Y) = \sum_{h=1}^{H} x^h(p, I^h, Y).$$

A necessary and sufficient condition for writing this aggregate demand function as a function of the aggregate income $I = \sum_{h=1}^{H} I^h$, instead of the individual incomes for arbitrary variations in income, is that all individuals have the same marginal propensity to demand out of income. Let us therefore assume that the individual demand functions are of the form

$$x_i^h = x_i^h(p, I^h, Y), \quad \frac{\partial x_i^h}{\partial I^h} = \frac{\partial x_i^k}{\partial I^k} = \beta_i; \quad h, k = 1,...,H, \quad i = 1,...,n.$$

We can then aggregate to

$$X_i = X_i(p, I, Y)$$

where

$$I = \sum_{h=1}^{H} I^h, \quad \frac{\partial X_i}{\partial I} = \beta_i.$$

Let us now define the aggregate expenditure function m as

$$m = \sum_{h=1}^{H} m^h.$$

Then

$$\frac{\partial m}{\partial p_i} = \sum_{h=1}^{H} \frac{\partial m^h}{\partial p_i}$$

and

$$\frac{\partial^2 m}{\partial p_i \partial p_j} = \sum_{h=1}^{H} \frac{\partial^2 m^h}{\partial p_i \partial p_j} = \sum_{h=1}^{H} \left\{ \frac{\partial x_i^h}{\partial I^h} \frac{\partial m^h}{\partial p_j} + \frac{\partial x_i^h}{\partial p_j} \right\} = \sum_{h=1}^{H} \left\{ \beta_i \frac{\partial m^h}{\partial p_j} + \frac{\partial x_i^h}{\partial p_j} \right\} = \beta_i \frac{\partial m}{\partial p_j} + \frac{\partial X_i}{\partial p_j}$$

and the aggregate expenditure function also satisfies the differential equations (4). Thus the condition for consistent aggregation of demand fumctions for arbitrary variations in individual incomes is also the conditions for consistent aggregation of expenditure functions. From now on it is assumed that this condition is fulfilled.

Samuelson [2] has shown that for a Pareto optimum in an economy with public goods, the following condition must be satisfied:

$$\sum_{h=1}^{H} \frac{\partial u^h}{\partial Y_k} \bigg/ \frac{\partial u^h}{\partial x_i^h} = MC_k^i$$

where MC_k^i is the social marginal opportunity cost of producing the public good Y_k in terms of production of the private good x_i.

Let the price of commodity 1 be p_1 and put

$$\delta_k^h = p_1 \frac{\partial u^h}{\partial Y_k} \bigg/ \frac{\partial u^h}{\partial x_1^h}.$$

δ_k^h will be interpreted as consumer h's demand price for the public good k. The purpose of this paper is to discuss one way of estimating the total demand price

$$\delta_k = \sum_{h=1}^{H} \delta_k^h$$

for the public good k.

Obviously, the total demand price will be a function of the price vector p, the incomes $I^1, \ldots, I^H$ and the vector of public goods supplied Y:

$$\delta_k = \sum_{h=1}^{H} \delta_k^h = \sum_{h=1}^{H} g_k^h(p, I^h, Y).$$

The expenditure function for consumer h is

$$m^h(p, \bar{u}^h, Y) = p^T x^{h+}.$$

If we differentiate the condition $u^h = \bar{u}^h$ with respect to Y_k we have

$$\sum_{i=1}^{n} u_i^h \frac{\partial x_i}{\partial Y_k} + \frac{\partial u^h}{\partial Y_k} = 0.$$

We also know that $\alpha^h u_i^h = p_i$ so that

$$\frac{\partial m^h}{\partial Y_k} = \sum_{i=1}^{n} p_i \frac{\partial x_i^{h+}}{\partial Y_k} = \alpha^h \sum_{i=1}^{n} u_i^h \frac{\partial x_i^{h+}}{\partial Y_k} = -\alpha^h \frac{\partial u^h}{\partial Y_k} = -p_1 \frac{\partial u^h}{\partial Y_k} \bigg/ \frac{\partial u^h}{\partial x_1^h} = -\delta_k^h.$$

The same relation holds for the total demand price

$$\frac{\partial m}{\partial Y_k} = -\delta_k.$$

The marginal willingness to pay or the demand price for the public good can therefore be estimated by estimating the expenditure function as a function of Y. But it was noted above that this is impossible if the only information we have consists of the demand functions for private goods. Thus further assumptions have to be added in order to solve our problem.

IV. Estimating Demand Prices for Public Goods

Note that assumption (1) is invariant for monotonic transformations of the utility function. In fact, let F be any monotonic increasing function. Then

$$\frac{\partial F(u^h(x_1, ..., 0, ..., x_n, Y))}{\partial Y_k} = F' \frac{\partial u^h(x_1, ..., 0, ..., x_n, Y)}{\partial Y_k}.$$

Note also that (1) is equivalent to

$$\frac{\partial m(p', \bar{u}, Y)}{\partial Y_k} = 0 \tag{5}$$

where p' is the price vector which causes a zero compensated demand for x_j, that is $x_j^+ = 0$ (for simplicity the superscript h is dropped).

Assume that there is a pair x_j, Y_k such that (1) is true. Let us now aggregate all other private goods to a composite good z with a price p_z and denote x_j by x. As the supply of all public goods except Y_k is going to be held constant, we can drop the corresponding variables and denote Y_k by Y.

The demand function for x can then be written

$$x = x(p_x, p_z, I, Y).$$

Consider the first equation in (3):

$$\frac{\partial^2 m}{\partial p_x^2} - \frac{\partial x(p_x, p_z, m, Y)}{\partial I} \frac{\partial m}{\partial p_x} - \frac{\partial x(p_x, p_z, m, Y)}{\partial p_x} = 0.$$

The general solution to this equation will have the form

$$m = \psi(p_x, p_z, Y, \varphi_1, \varphi_2) \tag{6}$$

where φ_1 and φ_2 are functions of p_z and Y.

The initial conditions are

$$\psi(\bar{p}_x, \bar{p}_z, \overline{Y}, \varphi_1, \varphi_2) = I$$

$$\frac{\partial \psi(\bar{p}_x, \bar{v}_z, \overline{Y}, \varphi_1, \varphi_2)}{\partial p_x} = x(\bar{p}_x, \bar{p}_z, I, \overline{Y})$$

where $\bar{p}_x$, $\bar{p}_z$, Y are the prices and supply of the public good in the initial situation. By solving the initial conditions the values of φ_1 and φ_2 in the initial situation can be obtained.

Note that the compensated demand function for x is given by

$$x^+ = \frac{\partial m}{\partial p_x} = \varkappa(p_x, p_z, Y, \varphi_1, \varphi_2).$$

Denote the inverse function by

$$p_x = h_x(x^+, p_z, Y, \varphi_1, \varphi_2).$$

Because the substitution effect is always negative, we know that this function always exists.

This function enables us to find those prices for which x^+ is zero:

$$p_x \geqslant p_x' = h_x(0, p_z, Y, \varphi_1, \varphi_2).$$

If we substitute this in (6) we obtain

$$m(p_x, p_z, Y) = m(p_x', p_z, Y) = \psi(h_x(0, p_z, Y, \psi_1, \varphi_2)) p_z, Y, \varphi_1, \varphi_2) \tag{7}$$

for $p_x \geqslant p_x'$.

Assumption (5) can now be applied. If assumption (1) or (5) holds, then the derivative of (7) with respect to Y is zero:

$$\frac{dm}{dY} = \frac{\partial \psi}{\partial p_x} \frac{\partial h_x}{\partial Y} + \frac{\partial \psi}{\partial Y} + \frac{\partial \psi}{\partial \varphi_1} \frac{\partial \varphi_1}{\partial Y} + \frac{\partial \psi}{\partial \varphi_2} \frac{\partial \varphi_2}{\partial Y} = 0. \tag{8}$$

This is a differential equation in the two unknown functions φ_1 and φ_2 .But (8) is an identity in p_z which implies that (8) can be differentiated with respect to p_z so as to obtain one more equation.

We thus have two differential equations in the two unknowns $\varphi_1(Y)$ and $\varphi_2(Y)$. By solving these two equations, m is determined wholly as function of p_x and Y. By utilizing standard theorems on the existence and uniqueness of solutions to differential equations we see that the solution obtained is the desired expenditure function.

This analysis has, however, been carried out under one implicit assumption, i.e. that the demand function for x has Y as an argument. If this is not the case, the differential equation (8) cannot be established. Later on it will be shown that this is an exception which is not likely to occur.

The following simple example will be used to clarify the procedure outlined above.

Suppose we have obtained estimates of the demand functions

$$x = I/2p_x - aY/2$$

$$z = I/2p_z + p_x aY/2p_z$$

where a is some positive constant.

The differential equations (3) become

$$\frac{\partial^2 m}{\partial p_x^2} - \frac{1}{2p_x}\frac{\partial m}{\partial p_x} + \frac{m}{2p_x^2} = 0$$

$$\frac{\partial^2 m}{\partial p_x \partial p_z} - \frac{1}{2p_z}\frac{\partial m}{\partial p_x} = \frac{aY}{2p_z}.$$

By substitution we see that

$$m = \varphi_1(p_z, Y)p_x^{\frac{1}{2}} + \varphi_2(p_z, Y)p_x$$

is a solution to the first equation. If this expression is substituted for m in the second equation, we obtain

$$\frac{1}{2}\frac{\partial \varphi_1}{\partial p_z}p_x^{-\frac{1}{2}} + \frac{\partial \varphi_2}{\partial p_z} - \frac{1}{4p_z}\varphi_1 p_x^{-\frac{1}{2}} - \frac{1}{2p_z}\varphi_2 = \frac{aY}{2p_z}.$$

Because both φ_1 and φ_2 are indepedent of p_x, this yields two equations

$$\frac{\partial \varphi_1}{\partial p_z} - \frac{1}{2p_z}\varphi_1 = 0$$

and

$$\frac{\partial \varphi_2}{\partial p_z} - \frac{1}{2p_z}\varphi_2 = \frac{aY}{2p_z}.$$

The solutions are

$$\varphi_1 = f(Y)p_z^{\frac{1}{2}}$$

$$\varphi_2 = C(Y)p_z^{\frac{1}{2}} - aY$$

where $f(Y)$ and $C(Y)$ are undetermined functions. However, if assumption (5) is applied, these functions can be determined.

The expenditure function is

$$m = fp_z^{\frac{1}{2}}p_x^{\frac{1}{2}} + Cp_z p_x^{\frac{1}{2}} - ap_z Y$$

and the compensated demand function for x is

$$x^+ = \frac{\partial m}{\partial p_x} = \frac{1}{2}fp_x^{-\frac{1}{2}}p_z^{\frac{1}{2}} + Cp_z^{\frac{1}{2}} - aY.$$

The compensated demand for x is zero when

$$p_z \geqslant p_z' = \frac{1}{4} \frac{f^2}{Y^2} \frac{p_z}{(a - p_z^{\frac{1}{2}} C/Y)^2}.$$

For $p_z \geqslant p_z'$ the expenditure function becomes

$$m(p_x, p_z, Y) = m(p_x', p_z, Y) = \frac{1}{2} \frac{f^2}{Y} \frac{p_z}{(a - p_z^{\frac{1}{2}} C/Y)}$$

$$+ \frac{1}{4} \frac{f^2}{Y} \frac{p_z}{(a - p_z^{\frac{1}{2}} C/Y)^2} \frac{C}{Y} p_z^{\frac{1}{2}} - \frac{1}{4} \frac{f^2}{Y} \frac{p_z}{(a - p_z^{\frac{1}{2}} C/Y)} a = \frac{1}{2} \frac{f^2}{Y} \frac{p_z}{(a - p_z^{\frac{1}{2}} C/Y)}.$$

If assumption (5) is true, then $m(p_x, p_z, Y)$ with $p_x \geqslant p_x'$, has to be independent of Y for all p_z. This can only be true if

$$f(Y) = A Y^{\frac{1}{2}}$$

and

$$C(Y) = B Y$$

This means that the expenditure function becomes

$$m = A \sqrt{p_x p_z} \, Y + (B \sqrt{p_z} - a) p_x Y.$$

The constants A and B can be determined from the initial conditions. If this is done, we find that

$$A = \frac{I + \bar{p}_x \overline{Y}}{\sqrt{\bar{p}_x \bar{p}_z} \, \overline{Y}}$$

$$B = 0$$

and

$$m = A \sqrt{p_x p_z} \, Y - a p_x Y.$$

It can easily be proved that A is the indirect utility function. Then

$$\frac{\partial A}{\partial I} = \frac{1}{\sqrt{\bar{p}_x \bar{p}_z} \, \overline{Y}} \doteq \lambda$$

$$\frac{\partial A}{\partial p_x} = \frac{1}{\sqrt{\bar{p}_x \bar{p}_z} \, \overline{Y}} \left[\frac{I}{2\bar{p}_x} - \frac{aY}{2} \right] = -\lambda x$$

$$\frac{\partial A}{\partial p_z} = \frac{1}{\sqrt{\bar{p}_x \bar{p}_z} \, \overline{Y}} \left[\frac{I}{2\bar{p}_z} + \frac{a\bar{p}_x \overline{Y}}{2\bar{p}_z} \right] = -\lambda z$$

(where λ is the Lagrange multiplier in (2)) can be used to solve for $\bar{p}_x$, $\bar{p}_z$ and I in terms of x, z, λ and $\overline{Y}$:

$$\bar{p}_x = \frac{1}{\lambda} z^{\frac{1}{2}} Y^{-\frac{1}{2}} (x + aY)^{-\frac{1}{2}}$$

$$\bar{p}_z = \frac{1}{\lambda} z^{-\frac{1}{2}} Y^{-\frac{1}{2}} (x + aY)^{\frac{1}{2}}$$

$$I = (2x + aY)\frac{1}{\lambda} z^{\frac{1}{2}} Y^{-\frac{1}{2}} (x + aY)^{-\frac{1}{2}}.$$

By substituting these expressions back into A, the original utility function is obtained (if the demand functions are those of a single individual):

$$u = 2(x + aY)^{\frac{1}{2}} z^{\frac{1}{2}} Y^{-\frac{1}{2}}.$$

The utility function has thus been derived on the basis of the demand functions for private goods by only assuming (5).

This example illustrates the technique of using assumption (5) to derive the expenditure function as a function of the public good Y. But the method fails if the differential equations (3) cannot be established. If the demand for x is a function of Y, then we have shown that (8) can be obtained by means of a routine calculation. The theorems on uniqueness of solutions to differential equations guarantee that the expenditure function ultimately derived is the correct one. But in the case where x is not a function of Y, (8) cannot be established and there is no possible way of deriving the expenditure function as a function of Y. But I will argue here that when the utility function is such that x does not depend on Y, assumption (5) is not likely to be realistic.

First, let us investigate the conditions under which the demand for a commodity does not depend on the amount of the public good.

Differentiating the optimality conditions (2) with respect to Y yields

$$u_{11} \frac{\partial x}{\partial Y} + u_{12} \frac{\partial z}{\partial Y} - p_x \frac{\partial \lambda}{\partial Y} = -u_{13}$$

$$u_{21} \frac{\partial x}{\partial Y} + u_{22} \frac{\partial z}{\partial Y} - p_z \frac{\partial \lambda}{\partial Y} = -u_{23}$$

$$-p_x \frac{\partial x}{\partial Y} - p_z \frac{\partial z}{\partial Y} = 0.$$

Let D be the determinant of coefficients in this equation system. Then the solutions for $\partial x / \partial Y$ and $\partial z / \partial Y$ are

$$\frac{\partial x}{\partial Y} = \frac{p_z}{D} (p_z u_{13} - p_x u_{23})$$

$$\frac{\partial z}{\partial Y} = -\frac{p_x}{D} (p_z u_{13} - p_x u_{23}).$$

The condition for $\partial x/\partial Y = \partial z/\partial Y = 0$ is

$$p_z u_{13} - p_x u_{23} = 0.$$

Due to (2) and because this is an identity, it can be written as

$$\frac{u_1}{u_{13}} = \frac{u_2}{u_{23}}$$

or

$$\frac{\partial}{\partial Y} \log u_1 = \frac{\partial}{\partial Y} \log u_2$$

which gives us

$$\log u_1 = \log (B(x, z)u_2)$$

where $B(x, z)$ is an arbitrary function of x and z.

This equation is equivalent to

$$u_1 - B(x, z) u_2 = 0$$

which is a partial differential equation of the first order, the characteristic of which is given by

$$dx + \frac{dz}{B(x,z)} = 0.$$

If B is differentiable, this equation has a solution which is given by

$$\varphi(x, z) = C$$

where C is an arbitrary integration constant.

The general solution to the partial differential equation can now be written

$$u = f(\varphi(x, z), Y)$$

where φ satisfies

$$\frac{\partial z}{\partial x} = -B(x, z).$$

If assumption (1) is now applied we find that

$$f_2(\varphi(0, z), Y) = 0.$$

Since this relation holds for all z, by differentiating with respect to z we find that

$$f_{21} \varphi_2(0, z) = 0.$$

$f_{21}=0$ is not a property which is invariant for monotone increasing transformations, so (if we want to stick to an ordinal approach)

$$\varphi_2(0, z) = 0$$

or

$$\frac{\partial}{\partial z} u(0, z, Y) = \frac{\partial}{\partial Y} u(0, z, Y) = 0.$$

This relation shows that if consumption of x is zero, then the consumer is indifferent to how much of the composite commodity z he consumes. This is a very strong statement about a certain complementariness between x and z. The assumption that the demand for private goods does not depend on the supply of the public good combined with (1) therefore yields a conclusion which is not likely to be realistic.

With respect to the type of analysis discussed here, cases where the demand for private goods does not depend on the supply of the public good can be disregarded with a high degree of confidence.

References

1. Karlin, S.: *Mathematical Methods and Theory in Games, Programming, and Economics*, Vol. I. Addison-Wesley, 1959.
2. Samuelson, P.: The pure theory of public expenditure. *Review of Economics and Statistics XXXVI*, no. 4, November, 1954.
3. Stevens, J.: Recreation benefits from water pollution control. *Water Resources Research*, Vol. 2, Second Quarter, 1966.

HEALTH AND AIR POLLUTION

The Effect of Occupation Mix

Lester B. Lave and Eugene P. Seskin*

Carnegie-Mellon University, Pittsburgh, Pennsylvania, USA

Summary

The 1960 mortality rates for 117 U.S. cities are analyzed to isolate the effect of air pollution. Socioeconomic factors such as race, income, population density, and age mix are accounted for. In previous studies we reported the effect of different functional forms and of adding additional explanatory variables for type of home heating equipment, home heating fuel, water heating fuel, air conditioning, and meterological factors. Occupation mix is now added and results in little change in the estimated air pollution coefficients. These health effects are translated into dollar terms.

Introduction

Evidence linking air pollution with ill health has been accumulating for some time.[1] Critics have argued that the studies are inadequate and that a causal link between air pollution and ill health is unproven.[2] Many studies have been criticized because they did not account for all of the possible factors which might cause the ill health. Thus, the estimated relation between air pollution and ill health might be spurious. The investigators have been criticized for assuming that the statistical association between air pollution and ill health is a causal relation, i.e., lowering the level of air pollution would lead to an increase in health levels.

These two objects are really synonymous. To estimate the relationship between air pollution and health, confounding factors must be accounted for. This means, other possible causes of illness must be controlled for or randomized in collecting the data, or they must be controlled for in the analysis. To show that an estimated association is causal rather than spurious, the relevant other factors must be accounted for.[3]

We have estimated the effect of air pollution on mortality rates for 117 Standard Metropolitan Statistical Areas (SMSAs) in 1960. The basic technique has been to make use of multiple regression analysis to control for con-

* This research was supported by a grant from Resources for the Future, Inc.
[1] Three recent literature reviews are Anderson (1967), Lave & Seskin (1970), and PHS (1970).
[2] See, for example, Battigelli (1968), Greenwald (1954), and WHO (1969).
[3] For a detailed discussion of this point see Lave & Seskin (1970c).

founding factors. The basic model postulates that the mortality rate (MR) in a city depends on the factors shown in equation (1):

$$MR = MR\ (A,\ F,\ G,\ N,\ S,\ E,\ Y,\ R,\ M,\ O,\ W,\ AP,\ e) \tag{1}$$

where A is a measure of the age distribution of the population, F is a measure of the sex distribution, G is a measure of various genetic factors, N is a measure of nutrition (both current and cumulative), S is a measure of smoking habits, E is a measure of exercise habits, Y is a measure of income, R is a measure of race, M is a measure of the quality and frequency of medical care, O is a measure of the occupation mix, W is a measure of climatological factors, AP is a measure of the air pollution, and e is an error term for omitted effects.

It is difficult, even conceptually, to measure some of these factors (such as genetic effects); it is difficult to get consistent data for others (such as medical care). We have attempted to introduce those factors which we can measure and have attempted to account for those we cannot measure, at least conceptually.

The goal of our research is the estimation of the quantitative effect of air pollution on health. We hope to be able to provide dollar measures of the benefit of pollution abatement as an input into a benefit–cost analysis to determine the "proper" quality of urban air. To achieve this goal, we must show that the estimated relation is not merely a sampling phenomenon and that it is not a spurious correlation. The large number of independent studies relating air pollution to ill health make it apparent that the relation is not a sampling phenomenon. Our attempts to control for the relevant possible confounding variables help to rule out the possibility that the association is a spurious correlation.

The crux of the problem is that laboratory evidence is not directly relevant to establishing the association. Neither is there a detailed physiological mechanism relating air pollution to ill health which would serve as a guide for empirical work. Instead, one must rely on observed data from natural experiments.

The nature of the controversy is similar to that surrounding the question of the relationship between cigarette smoking and ill health. Many hypotheses were proposed as to the "true" cause of the association: suggestions included genotype, nervous tension, and even air pollution being the true causes of lung cancer. Indeed, all of the skeptics are still not convinced of the relation.

In this paper we continue our attempt to account for the factors that have been suggested to be the "true" cause of the measured increase in mortality which we have associated with air pollution. We investigate how measures of occupation mix affect the mortality rate and what the effect on the air pollution estimates is of adding the occupation variables. This work supports earlier reports in which we added climate measures of the city, types of home heating equipment, home heating fuel, and water heating fuel of the region, and the presence of air-conditioning in the area (Lave & Seskin, 1970a). We have also looked at mortality rates which are specific to various diseases and to groups classified by age, sex, and race (Lave & Seskin, 1970c).

Occupation mix is relevant to the mortality rate for a number of reasons. Certain occupations, such as coal mining, are known to have high accident rates. Other occupations, such as making coke, expose the worker to a highly polluted atmosphere. If air pollution were associated with either high occupational exposures or high accident rates, the coefficients estimated previously would be biased.

The data[1]

We collected data on 117 SMSAs. Air pollution data were reported by the U.S. Public Health Service. Suspended particulates and total sulfates were measured for 26 biweekly periods in micrograms per cubic meter ($\mu g/m^3$). Observations were collected on the biweekly minimum and maximum readings and the annual arithmetic mean.[2] There are a number of difficulties with these data. The measuring instruments change over time and across cities and some instruments have little reliability. In addition, the data are generally for a single point in a vast geographical area. Since pollution concentration varies greatly with the terrain, it is a heroic assumption to regard the figures as representative of an entire SMSA in making comparisons across areas.

Mortality data are reported in *Vital Statistics of the United States*. These include the total death rate, a breakdown of the total death rate and various categories of infant death rates (as a ratio to live births), and a breakdown of the total death rate by individual disease categories. One problem with the infant death rates is that a classification such as fetal deaths will not be reported uniformly well across all areas. A difficulty with the disease specific rates is the accuracy of the diagnosis of the cause of death, since few are verified by autopsy and not all physicians determine the cause of death with equal skill.

Finally, the socioeconomic data (including occupational information) were taken from the 1960 census as reported in the *County and City Data Book*.

The variables which we use along with their means and standard deviations are reported in the footnote to Table 1.

We again stress the fact that our measures of air pollution are quite imperfect. They are for a single point in the city, while there is known to be major variation in the different areas of the cities. Thus, it seems likely that occupation mix is acting as a surrogate, in part, for unmeasured air pollution. Industries which subject their workers to a highly polluted working

---

[1] A vast number of qualifications and problems are alluded to in this report. These are explored in more detail in Lave (1971).
[2] For a few areas it was necessary to use an adjoining year, i.e., 1959, because of incomplete data.

environment have a marked effect on the general amount of air pollution in the city. Therefore, we would expect that these occupation variables may take away some of the apparent significance of the air pollution variables.

Method

In a previous investigation (Lave & Seskin, 1970b) we determined the "best" set of regressions utilizing a step-wise regression technique. The explanatory variables included socioeconomic characteristics of the city as well as measures of air pollution. In the present exposition, we add a set of variables representing the occupational mix of the community. We proceed to examine each basic regression before and after the inclusion of these occupation variables. We should mention at this point that in the discussion which follows primary concern will be given to the effect of the occupation variables on the pollution and socioeconomic variables with secondary emphasis on the significant vocational variables themselves.

Results: total deaths

The most accurate and complete mortality data are for total deaths. They are accurate since there is little or no reporting problem. However, people migrate in response to health problems and also take steps to lessen their exposure to air pollution by moving to less polluted areas within the city or installing devices to filter out pollution.

While the total death rate has the virtue of being quite accurate, it has the drawback of being so aggregate that many interesting questions cannot be considered. An analysis of age specific death rates is needed to look at the effect of pollution on life expectancy; disease specific death rates would shed more light on the physiological mechanisms involved. These various death rates tend to give independent confirmation of the effect of air pollution on health.

The first regression in the Table at the end of the paper is for the total mortality rate and is reproduced in equation (2):

$$MR_i = 19.607 + \underset{(2.53)}{.041} \text{ Mean } P + \underset{(3.18)}{.071} \text{ Min } S$$
$$+ \underset{(1.67)}{.008} P/M^2 + \underset{(5.81)}{.041}\% \text{ NW} + \underset{(18.94)}{.687}\% \geqslant 65 + e_i \qquad (2)$$

where "Mean P" is the arithmetic mean of the 26 suspended particulate readings, "Min S" is the smallest of the 26 biweekly sulfate readings, "P/M^2" is the population density in the SMSA, "% NW" is the percentage of the SMSA population who are nonwhite, "% $\geqslant 65$" is the percentage of the SMSA population who are 65 and older, and "e" is an error term. This regression explains variations in the total mortality rate across 117 SMSAs extremely well, since 82.7 % of the variation is explained ($R^2 = 0.827$). Each of the coefficients except population density is extremely significant, as shown by the t

Table 1. *Regression analysis of mortality rates*

	Total 1	Total 2	<1 Year 3	<1 Year 4	<28 Days 5	<28 Days 6	Fetal 7	Fetal 8	TB 9	TB 10	Cancers 11	Cancers 12
R^2	0.827	0.917	0.537	0.648	0.271	0.421	0.426	0.532	0.234	0.434	0.879	0.926
Constant	19.607	80.368	185.802	448.346	149.428	274.746	93.852	180.946	-0.169	0.499	3.247	-0.136
Pollution:												
Min P			0.365 (2.82)	0.353 (2.65)								
Max P												
Mean P	0.041 (2.53)	0.027 (2.04)			0.083 (1.62)	0.072 (1.31)						
Min S	0.071 (3.18)	0.033 (1.66)			0.120 (1.82)	0.124 (1.59)					0.009 (2.06)	0.008 (1.76)
Max S												
Mean S							0.141 (2.67)	0.045 (0.73)			0.006 (1.96)	-0.001 (-0.47)
Socioeconomic:												
P/M^2	0.001 (1.67)	-0.0004 (-0.74)					0.003 (1.61)	-0.001 (-0.55)	0.00005 (2.65)	-0.00001 (-0.62)	0.0003 (3.61)	0.00005 (0.50)
N·W	0.041 (5.81)	0.025 (3.39)	0.186 (6.52)	0.143 (4.16)	0.098 (4.04)	0.067 (2.19)	0.161 (5.33)	0.126 (3.32)			0.005 (3.35)	0.005 (2.82)
≥65	0.687 (18.94)	0.613 (19.43)							0.002 (1.59)	0.001 (1.08)	0.132 (20.21)	0.126 (20.06)
Poor			0.157 (3.38)	0.128 (1.72)	0.056 (1.45)	0.055 (0.84)	0.125 (2.49)	0.167 (2.06)	0.002 (4.55)	0.001 (1.38)	-0.010 (-4.47)	-0.012 (3.76)

Occupation:					
% Unem.	0.114 (2.46)	-0.017 (-0.08)	0.124 (0.49)	0.005 (2.56)	0.027 (2.72)
% Male	-0.032 (-1.15)	-0.091 (-0.77)	-0.123 (-0.86)	-0.001 (-0.41)	0.008 (1.35)
% Agr.	-55.926 (-1.84)	-243.601 (-1.91)	227.273 (1.40)	-1.806 (-1.30)	-3.805 (-0.60)
% Con.	-128.367 (-2.10)	-19.190 (-0.07)	-42.207 (-0.13)	2.831 (1.01)	15.058 (-1.20)
% Mfr. (D)	-15.471 (-0.94)	-46.988 (-0.67)	80.083 (0.91)	-0.078 (-0.10)	-1.149 (-0.34)
% Mfr. (ND)	8.112 (0.48)	-133.761 (-1.88)	15.374 (0.17)	0.184 (0.24)	4.469 (1.28)
% Trans.	71.793 (2.01)	-96.756 (-0.65)	64.542 (0.34)	0.216 (0.13)	4.512 (0.62)
% Trade	-55.353 (-1.64)	111.660 (0.73)	-91.646 (-0.48)	-1.305 (-0.79)	-5.867 (-0.78)
% Fin.	59.373 (1.00)	66.348 (0.27)	-103.651 (-0.33)	3.060 (1.13)	20.624 (1.70)
% Educ.	-19.410 (-0.48)	-399.291 (-2.26)	-291.417 (-1.34)	0.873 (0.46)	-0.260 (-0.03)
% Pub.	23.136 (1.14)	-46.181 (-0.54)	94.994 (0.88)	1.378 (1.48)	5.650 (1.35)
% W-C	-0.052 (-2.72)	-0.059 (-0.65)	-0.026 (-0.23)	-0.001 (-1.47)	-0.005 (-1.15)
% P. Trans.	0.019 (1.67)	0.008 (0.20)	0.169 (2.86)	0.001 (3.06)	0.008 (3.40)

Table 1. (Continued)

	Buccal C.		Stomach C.		Resp. C.		Breast C.		Asthma		Cardiovascular	
	13	14	15	16	17	18	19	20	21	22	23	24
R^2	0.352	0.445	0.775	0.866	0.606	0.698	0.705	0.773	0.083	0.222	0.827	0.915
Constant	-0.125	0.445	0.499	5.743	0.277	-5.251	0.556	1.023	0.062	0.737	1.214	21.888
Pollution:												
Min P	0.001	0.001										
	(1.12)	(1.16)										
Max P												
Mean P									0.001	0.001		
									(2.64)	(2.84)		
Min S			0.006	0.003							0.058	0.021
			(2.48)	(1.42)							(3.14)	(1.32)
Max S			0.001	0.0002			0.0003	0.0001				
			(2.29)	(0.41)			(1.70)	(0.71)				
Mean S					0.001	0.0005					0.028	0.010
					(1.51)	(0.51)					(2.58)	(1.00)
Socioeconomic:												
P/M^2	0.00002	0.00001	0.0002	0.00003	0.0001	0.0001	0.00004	-0.000001				
	(1.85)	(0.67)	(4.03)	(0.45)	(3.36)	(2.90)	(2.69)	(-0.07)				
N-W	0.0004	0.0003			0.002	0.003	0.001	0.0003			0.18	0.015
	(3.31)	(1.56)			(3.76)	(4.22)	(2.72)	(0.95)			(3.45)	(2.69)
≥ 65	0.004	0.004	0.048	0.044	0.021	0.023	0.012	0.010	0.001	0.001	0.469	0.448
	(6.98)	(4.83)	(13.85)	(13.26)	(9.91)	(9.79)	(10.59)	(8.53)	(2.00)	(1.14)	(17.97)	(19.33)
Poor			-0.003	-0.006	-0.002	-0.002	-0.002	-0.002	0.0002	-0.0004		
			(-2.72)	(-3.29)	(-2.19)	(-1.98)	(-6.32)	(-3.92)	(1.23)	(-1.10)		

Occupation:						
% Unem.	0.002 (1.94)	0.019 (3.29)	0.002 (0.60)	0.002 (1.01)	0.001 (0.50)	0.059 (1.69)
% Male	−0.001 (−1.60)	−0.002 (−0.52)	0.006 (2.76)	−0.001 (−1.05)	−0.0004 (−0.48)	−0.001 (−0.05)
% Agr.	−0.543 (−0.72)	−3.418 (−0.89)	−0.621 (−0.26)	2.714 (2.26)	−0.437 (−0.56)	−37.888 (−1.63)
% Con.	−0.820 (−0.54)	−11.479 (−1.47)	3.831 (0.81)	0.620 (0.25)	1.239 (0.79)	−84.846 (−1.84)
% Mfr. (D)	0.071 (0.17)	−2.658 (−1.27)	0.718 (0.56)	0.576 (0.88)	−0.167 (−0.39)	−1.455 (−0.12)
% Mfr. (ND)	0.244 (0.58)	−0.678 (−0.33)	1.777 (1.36)	0.973 (1.46)	−0.115 (−0.27)	23.137 (1.80)
% Trans.	−0.854 (−0.97)	−0.829 (−0.18)	5.511 (2.03)	−1.354 (−0.96)	0.083 (0.09)	37.738 (1.41)
% Trade	0.751 (0.89)	−9.056 (−2.00)	3.193 (1.14)	0.239 (0.17)	0.425 (0.46)	−35.957 (−1.42)
% Fin.	0.045 (0.03)	10.612 (1.43)	5.058 (1.12)	1.915 (0.82)	3.813 (2.54)	26.482 (0.60)
% Educ.	0.817 (0.82)	1.961 (0.38)	0.284 (0.09)	−0.845 (−0.52)	−0.143 (−0.13)	10.177 (0.34)
% Pub.	−0.091 (−0.18)	−1.337 (−0.52)	2.635 (1.69)	0.870 (1.08)	0.074 (0.14)	24.912 (1.62)
% W-C	0.000004 (0.00)	−0.004 (−1.53)	−0.001 (−0.67)	−0.0001 (−0.12)	−0.001 (−2.54)	−0.033 (−2.22)
% P. Trans.	0.0002 (0.84)	−0.005 (3.51)	−0.0002 (−0.19)	0.001 (3.25)	−0.0002 (−1.11)	−0.008 (1.14)

Table 1. (*Continued*)

	Heart.		Endocarditis		Hypertensive		Influenza		Pneumonia		Bronchitis	
	25	26	27	28	29	30	31	32	33	34	35	36
R^2	0.538	0.641	0.395	0.468	0.431	0.598	0.339	0.477	0.149	0.308	0.077	0.203·
Constant	1.643	56.098	-2.600	0.923	-0.661	-1.786	0.014	-1.386	1.481	3.687	0.197	1.015
Pollution:												
Min P												
Max P			0.002 (1.85)	0.002 (1.31)					0.003 (0.74)	0.002 (0.36)	0.0001 (1.05)	0.0002 (2.12)
Mean P												
Min S	0.066 (2.34)	0.071 (2.25)	0.008 (1.53)	0.007 (1.10)	0.006 (1.75)	0.002 (0.59)						
Max S	0.013 (1.97)	0.002 (0.30)			0.003 (3.15)	0.003 (2.87)						
Mean S			0.005 (1.38)	0.008 (1.68)			0.001 (1.23)	0.0003 (0.57)				
Socioeconomic:												
P/M^2	0.001 (1.69)	-0.00004 (-0.05)	0.0003 (2.30)	0.0002 (1.62)			-0.00002 (-1.17)	-0.00002 (-1.16)	0.0001 (1.68)	0.000002 (0.02)		
N-W					0.009 (8.30)	0.009 (6.72)	-0.001 (-1.90)	-0.0004 (-1.22)	0.001 (1.47)	-0.0004 (-0.37)		
≥65	0.314 (8.27)	0.333 (8.21)	0.032 (4.70)	0.027 (3.46)	0.026 (4.69)	0.024 (4.07)	-0.002 (-1.29)	-0.001 (-0.93)	0.015 (3.54)	0.006 (1.27)	0.001 (1.63)	0.0004 (0.74)
Poor			0.006 (2.59)	0.001 (0.17)			0.003 (6.22)	0.004 (5.57)			-0.0003 (-2.01)	-0.001 (-3.13)

Occupation:						
% Unem.	-0.105 (-1.43)	0.026 (1.89)	0.022 (2.52)	-0.001 (-0.61)	0.006 (0.76)	0.042 (1.70)
% Male	0.033 (0.78)	-0.008 (-0.93)	-0.003 (-0.59)	0.002 (1.63)	-0.006 (-1.29)	-0.002 (-2.66)
% Agr.	-63.221 (-1.30)	2.505 (0.28)	1.890 (0.32)	-1.188 (-0.89)	4.242 (0.85)	-0.033 (-0.05)
% Con.	-155.719 (-1.58)	11.874 (0.66)	-2.975 (-0.25)	-5.438 (-2.04)	-3.515 (-0.35)	1.414 (1.05)
% Mfr. (D)	-78.472 (-2.97)	1.279 (0.26)	3.573 (1.14)	0.081 (0.11)	0.450 (0.17)	0.520 (1.43)
% Mfr. (ND)	-42.642 (-1.64)	4.520 (0.93)	8.400 (2.61)	0.197 (0.27)	2.909 (1.05)	-0.034 (-0.09)
% Trans.	-126.547 (-2.21)	13.867 (1.33)	12.040 (1.78)	3.307 (2.16)	2.349 (0.41)	0.969 (1.26)
% Trade	-180.073 (-3.33)	-0.846 (-0.08)	-1.390 (-0.22)	-1.933 (-1.22)	7.241 (1.31)	1.801 (2.29)
% Fin.	-97.109 (-1.02)	2.081 (0.12)	8.389 (0.75)	5.027 (1.97)	-1.236 (-0.13)	1.037 (0.81)
% Educ.	-164.612 (-2.56)	6.886 (0.57)	11.401 (1.49)	1.808 (1.02)	-10.760 (-1.64)	1.627 (1.82)
% Pub.	-81.124 (-2.48)	7.895 (1.33)	11.380 (2.93)	-0.289 (-0.33)	-1.279 (-0.38)	0.460 (1.04)
% W-C	0.032 (1.11)	-0.003 (-0.50)	-0.002 (-0.63)	0.0003 (0.36)	0.002 (0.73)	-0.001 (-1.97)
% P. Trans.	0.020 (1.21)	-0.001 (-0.45)	-0.0005 (-0.25)	-0.001 (-1.49)	0.003 (1.40)	0.00003 (0.16)

Notes to Table.

Columns 1 and 2 report two regression explaining the Total mortality rate. Column 2 adds occupation variables to the air pollution and socioeconomic explanatory variables of Column 1. R^2 is the coefficient of determination; the air pollution and socioeconomic variables explain 82.7 % of the variation in the Total mortality rate across the 117 cities. The estimated coefficient of Mean P (the arithmetic mean of suspended particulates) is 0.041; the t statistic is 2.53, indicating the coefficient is significantly larger than zero at a 95 % confidence level.

Variables used in the analysis

	Mean	Standard Deviation
Air pollution		
Suspended particulates (μg/m^3)		
Minimum reading for a biweekly period (1960)	45.47	18.57
Maximum reading for a biweekly period	286.36	132.07
Arithmetic mean (annual)	118.14	40.94
Total sulfates (μg/m$^3 \times 10$)		
Minimum reading for a biweekly period	47.24	31.28
Maximum reading for a biweekly period	228.39	124.41
Arithmetic mean (annual)	99.65	52.88
Mortality		
Total death rate (per 10 000)	91.26	15.33
Infant death rate (per 10 000 live births)		
< 1 year	254.03	36.44
< 28 days	187.29	24.52
Fetal	153.15	34.35
Diseases (per 10 000)		
Tuberculosis of respiratory system	0.55	0.27
Malignant neoplasms, including neoplasms of lymphatic and hemopoietic tissues	14.34	3.30
Malignant neoplasm of buccal cavity and pharynx	0.34	0.15
Malignant neoplasm of digestive organs and pertoneum, not specified as secondary	4.72	1.51
Malignant neoplasm of respiratory system, not specified as secondary	2.22	0.61
Malignant neoplasm of breast	1.26	0.36
Asthma	0.29	0.13
Diseases of cardiovascular system	48.25	11.47
Diseases of heart	34.93	11.85
Nonrheumatic chronic endocarditis and other myocardial degeneration	2.80	1.75
Hypertensive heart diseases	3.53	1.34
Influenza	0.34	0.26
Pneumonia, except of newborn	3.13	0.87
Bronchitis	0.23	0.11
Socioeconomic		
Persons per square mile	756.15	1 370.54

	Mean	Standard Deviation
% non-whites in population (× 10)	125.06	103.98
% population ⩾ 65 (× 10)	83.93	21.21
% families with incomes < $3 000 (× 10)	180.85	65.53
Occupation		
% Unemployed (× 10)	49.92	15.58
% Male (× 10)	656.20	30.59
% in Agriculture (/100)	0.02	0.02
% in Construction (/100)	0.06	0.01
% in Manufacturing (durable goods) (/100)	0.14	0.10
% in Manufacturing (nondurable goods) (/100)	0.12	0.07
% in Transportation, communication, and other public utilities (/100)	0.07	0.02
% in Wholesale and retail trade (/100)	0.18	0.02
% in Finance, insurance and real estate (/100)	0.04	0.01
% in Educational services (/100)	0.05	0.02
% in Public administration (/100)	0.05	0.04
% in Whitecollar occupations (× 10)	437.02	54.48
% Using public transportation to work (× 10)	110.17	72.13

statistics below the coefficients. As expected, increases in each of the variables, other variables held constant, would lead to an increase in the total mortality rate.

The percentage of older people is the most important variable in equation (2). A one percentage point increase in the proportion multiplied by ten (× 10) of people 65 and older (raising the mean from 83.93 to 93.93) is estimated to raise the total death rate 6.87 per 10 000 from a mean of 91.26 to 98.13). Increasing nonwhites in the population (× 10), by 1 percentage point (raising the mean from 125.06 to 135.06), is estimated to raise the total death rate by 0.41 per 10 000. If air pollution worsened and either the minimum sulfate level or mean particulate level rose by 1 $\mu g/m^3$, the total death rate would rise by either 0.71 or 0.04, respectively.

One can interpret the results of regression 1 in a slightly different fashion than the above explanation. A 10 % decrease in the mean level of particulate pollution would lead to a 0.53 % decrease in the total death rate (a decrease of 11.81 $\mu g/m^3$ in the annual mean level of particulate pollution would lead to a decrease of 0.48 deaths per 10 000). A 10 % decrease in the minimum level of sulfate pollution is estimated to reduce the total mortality rate by 0.37 % (a decrease of 0.47 $\mu g/m^3$ in the biweekly minimum sulfate reading would lead to a decrease of 0.33 deaths per 10 000). A 10 % decrease in the population density would lead to a 0.08 percent decrease in the total death rate (a decrease of 75.62 P/M^2 would reduce total deaths by 0.08 per 10 000). A 10 % decrease in the percentage of nonwhites in the population is estimated to lead to a reduction of 0.56 percent in the total death rate (a 1.25 percentage point decrease in the proportion of nonwhites would reduce total deaths by 0.51 per 10 000). A 10 % decrease in the percentage of people 65 and older in the population is estimated to lead to a reduction of 6.32 % in the total mortality rate (a 0.84 percentage decrease in the proportion of older people would reduce total deaths by 5.77 per 10 000).

The employment variables are added in regression 2. When comparing the pollution variables in the two regressions, one notices that although the coefficients and the *t* statistics are decreased, the pollution variables remain significant statistically. The coefficients of the socioeconomic variables are also diminished slightly, but only population density becomes insignificant statistically.

Looking at the significant occupation variables, we note that the coefficient of unemployment has the expected sign, i.e., an increase in the proportion of the working force unemployed is associated with an increase in the total mortality rate. Increases in the proportion of agricultural and construction workers are associated with decreases in the death rate. Such employment is carried out largely in a rural setting subject to less pollution. As the percentage of people working in transportation, communication, and other public utilities increases, the total death rate is estimated to increase. This is quite plausible since we would expect these workers to be exposed to relatively high levels of pollution. Increases in the proportion of the labor force employed in wholesale and retail trade is associated with decrease in the mortality rate (probably reflecting the indoor nature of the job). If the types of occupations are aggregated, then we find that as the percentage of white-collar workers increases, the total death rate decreases. Finally, as the proportion of people utilizing public transportation to and from work increases, the total mortality rate increases.

Infant deaths

Data were available for three classes of infant mortality rates (per 10 000 live births): the rate for infants under one year, the rate for infants under 28 days, and the rate for stillborns. The first two of these rates are presumably quite accurate in that both births and deaths are tabulated correctly. The fetal mortality rate, however, is subject to underreporting in certain areas, particularly if the woman is not under a physician's care.

The under one year death rate is analyzed in regression 3. Just over half (54 %) of the variation in the mortality rate is explained across the 117 SMSAs. The coefficient of the pollution variable indicates that a decrease of 4.55 $\mu g/m^3$ (10 %) in the biweekly minimum level of particulate pollution would lead to a decrease of 1.64 (0.65 %) death per 10 000 live births in the under-one-year category.

When the occupation variables are added to the above regression (in regression 4), the pollution and socioeconomic variables remain significant although their magnitudes fall slightly. Again, agricultural employment is associated with decreases in the death rate. In addition, as the proportion of workers employed in the manufacturing of nondurable goods increases, the death rate in this category decreases. Educational employment also seems related to lower death rates of infants under one year. Perhaps women so employed prior to pregnancy are subject to less strenuous activity (and less pollution) than women in other occupations. Finally, we see that the composite category of white-collar workers is again associated with lower death rates.

Regression 5 explains the under-28-day infant mortality rate in terms of mean particulates, minimum sulfates, nonwhites, and poor families. About 27 % of the variation across SMSAs is explained, indicating that the explanatory power of the regression, while statistically significant, is less good than the previous one. According to the estimates, a decrease of 11.81 $\mu g/m^3$ (10 %) in the annual mean level of particulate pollution would lead to a decrease of 0.98 (0.52 %) deaths per 10 000 live births in the under-28-day infant group; a decrease of 0.47 $\mu g/m^3$ (10 %) in the biweekly minimum level of sulfates would lead to a decrease of 0.56 (0.30 %) death per 10 000 live births in this category.

The occupation variables are added in regression 6. Comparing the pollution and socioeconomic variables with the previous results, we again see decreases in the magnitudes of the coefficients (except for minimum sulfates). The t statistics of all the variables are affected adversely with the pollution variables approaching statistical significance and the poor variable becoming insignificant. The significant occupational variables are similar to the under one year category described above (the white-collar classification is not significant in this case), although the magnitudes of the coefficients and the t statistics are in general diminished.

The fetal death rate is analyzed in regression 7. A greater percentage of the variation in fetal deaths is explained (compared to the under-28-day death rate). The important pollution variable for fetal deaths appears to be the annual mean level of sulfates. According to the estimated coefficient, a decrease of 1.00 $\mu g/m^3$ (10 %) in the mean level would lead to a decrease of 1.41 (0.92 %) fetal deaths per 10 000 live births.

With the addition of the occupation variables, the pollution variable used in explaining the variation in the fetal death rate becomes insignificant (regression 8). In addition, population density loses significance. The only occupation variable which is significant statistically represents the utilization of public transportation by workers. The coefficient indicates that as the percentage of individuals using such public means increases, the death rate in this category increases.

Mortality rates for specific diseases

In tabulating death rates, either for all deaths or for only infant deaths, there are difficulties with estimating the population at risk, and difficulties if people migrate from city to city, but there is little or no trouble in determining the number of people who died. Death rates for specific diseases are subject to many additional problems. Diagnosis is imperfect and not all physicians take the same care in ascertaining the cause of death. Another difficulty is that the death rate for many diseases is so small that there is a great deal of sampling variation from year to year in a small-sized city. This sampling variation is likely to be important and we conjectured that the regressions would have higher explanatory power for the more prevalent causes of death. The 14 categories of disease which we analyzed are shown in the Table.

Tuberculosis is analyzed in regression 9. Since TB is a communicable disease, one would expect that population density would be a significant variable, as it is. TB is a disease of the poor, as shown by the regression. We would also expect that TB would be more prevalent among the old, as is shown in the regression. The mean level of particulate pollution is a significant explanatory factor for the TB death rate.

When the occupation variables are added (regression 10), the pollution variable remains quite significant despite a reduction in magnitude. The socioeconomic variables become insignificant statistically. Two occupational variables are significant. The first indicates that the percentage of unemployment is associated positively with the tubercular death rate. The second suggests that as the proportion of workers traveling to and from their jobs on public transportation increases, deaths from tuberculosis increase.

Death from all cancers is analyzed in regression 11; the explanatory power of the regression is quite large ($R^2 = 0.879$). Both the minimum and mean level of sulfate pollution are significant statistically. Unlike TB, cancer is more prevalent among people with higher incomes. In addition to the income variable, all of the other socioeconomic variables are important; population density, the percentage of nonwhites, and the percentage of people 65 and older are all positively related to the cancer death rate.

The occupation variables are entered in the next regression (12). The minimum level of sulfate pollution remains significant, while the mean level becomes insignificant. Of the socioeconomic variables, only population density loses statistical significance. The two occupation variables which were significant in explaining tuberculosis mortality are also significant in explaining deaths from all types of cancer. In addition, the percentage of people employed in finance, insurance and real estate is related positively to the cancer death rate.

Regression 13 presents the analysis of cancer of the buccal cavity and pharynx. As one might expect, each of the individual cancer categories is less well explained than the total category. None of the pollution variables is a significant explanatory variable, although the minimum level of particulates approaches significance. Population density, the percentage of nonwhites, and the percentage of people 65 and older are positively related to the incidence of this cancer.

Occupation variables are added to this regression in 14. The pollution variable is essentially unaffected, while the socioeconomic variables lose some significance (population density becomes insignificant statistically). Again, the unemployment rate is associated positively with this cancer rate. In addition, the percentage of males comprising the labor force is associated negatively with this death rate.

Cancer of the stomach is analyzed in regression 15. This type of cancer appears to be closely related to air pollution since both the minimum and maximum levels of sulfates are extremely significant. Population density and the number of people 65 and older are related positively to the incidence of stomach cancer, while the percentage of poor families is related negatively.

With the addition of the occupation variables (regression 16), the two pollution variables lose significance. Again, population density becomes insignificant statistically. The percentage of old people is unaffected, while the poor variable actually becomes more significant. As in the previous cases, increased unemployment is related significantly to an increased death rate, and the aggregate white-collar category is inversely related to the death rate. In addition, employment in wholesale and retail trade is related negatively to this class of cancer deaths, while utilization of a public means of transportation by workers is related positively to these deaths.

Regression 17 reports the results for cancer of the respiratory system. The only pollution variable which approaches significance is the mean level of sulfate pollution. All socioeconomic variables are significant and related in the same manner as for all cancers.

The occupation variables are added in the next regression (18). Mean sulfates become insignificant statistically, while the socioeconomic variables all remain significant. For this type of cancer, as the percentage of males in the labor force increases, the death rate increases. The other significant occupation variables indicate that as the percentage of employment in transportation, communication and public utilities or in public administration increases, the death rate in this category of cancer increases.

Breast cancer follows a pattern similar to the other cancers discussed above. The only pollution variable which is significant in regression 19 is the maximum sulfate level. The socioeconomic variables, again have the same signs as for all cancers.

Regression 20 exhibits the results when the occupation variables are added. Of the original variables, only the proportion of older people and the proportion of poor people remain significant explanatory variables. Contrary to results above, agricultural employment is related positively to this mortality rate. In addition, as the number of women using public transportation to work increases, the incidence of breast cancer is estimated to increase.

The percentage of variation across SMSAs of deaths due to asthma explained by regression 21 is quite small ($R^2 = 0.083$). This is largely due to sampling variation. The mean particulate level is quite significant. The only socioeconomic variable which is significant is the percentage of people 65 and older.

In the presence of the occupation variables (regression 22) the pollution variable is enhanced slightly, while the two socioeconomic variables become insignificant statistically. The aggregate category of white-collar employment is associated negatively with the asthmatic death rate. We offer no explanation for the positive association found between employment in finance, insurance and real estate and the asthma mortality rate.

Deaths from diseases of the cardiovascular system are explained quite well by pollution and socioeconomic variables (regression 23). Both the minimum and mean sulfate pollution are highly significant variables. The percentage of nonwhites in the population and the percentage of old people in the population are significant socioeconomic variables.

The variables representing occupational mix are added in regression 24. The two pollution variables lose statistical significance, while the socioeconomic variables maintain their significance. Again, unemployment adversely affects the death rate. Agricultural and construction areas of employment are related negatively to the cardiovascular mortality rate. In addition, as the proportion of individuals employed in the manufacturing of nondurable goods or public administration increases, the cardiovascular mortality rate is estimated to increase. Finally, we again note that the aggregate category of white-collar workers is related inversely with this death rate.

The analysis for diseases of the heart is shown in regression 25. Both the minimum and maximum sulfate pollution variables are significant. In addition, population density and the percentage of old people are important socioeconomic variables.

The maximum sulfate variable loses statistical significance when the occupation variables are included in regression 26. The minimum sulfate level remains quite significant. In addition, population density again loses significance. The following occupational categories are significant and all related negatively with the death rate from heart disease: construction, manufacturing (both durable and nondurable goods), transportation, communication, and other public utilities, wholesale and retail trade, education, and public administration.

Mortality from nonrheumatic chronic endocarditis and other myocardial degeneration is explained in regression 27. The maximum particulate pollution is related significantly to the death rate in this category, while both the minimum and mean level of sulfate pollution approach statistical significance. Population density, the percentage of old people in the population, and the percentage of poor people in the population are important socio-economic variables.

In the presence of the occupation variables (regression 28), the significance of the pollution variables is altered, leaving the mean sulfate level the most important measure of pollution. The socioeconomic variables lose some significance with the poor variable becoming insignificant statistically. The only occupation variable which is significant is the percent of unemployment in the labor force. Its interpretation is the same as above.

The analysis of death due to hypertensive heart disease is shown in regression 29. Both the minimum and maximum sulfate pollution variables are significant. In addition, the percentage of nonwhites and the percentage of older people in the population are important socioeconomic variables.

The main result to note with the addition of the occupation variables in regression 30 is that the minimum sulfate variable becomes insignificant statistically. The other variables are not. affected greatly. For this death rate, the percentage of unemployment as well as the following classifications are associated positively: manufacturing (nondurable goods), transportation, communication, and other public utilities, and public administration.

Deaths from influenza are examined in regression 31. The only pollution variable which approaches significance is the mean sulfate level. The socioeconomic variable of most importance is the percentage of poor people in the population; influenza appears to be a disease of the poor.

The addition of the occupation variables (regression 32) does not alter substantially these results. Of the significant occupational categories, construction work is related negatively to influenza mortality while transportation, communication, and other public utilities as well as finance, insurance and real estate are related positively to the death rate. One explanation for this might lie in the communicable nature of the disease.

Only 15 % of the variation across SMSAs of deaths from pneumonia is explained by regression 33. Again, this is probably due to sampling variation. No pollution variable is significant, and in this case the socioeconomic variable of most importance appears to be the number of people 65 and older in the population.

The significance of these variables is further reduced with the addition of the occupation variables in regression 34. None of the employment characteristics are significant statistically (the educational category approaches significance).

The last disease we analyze is bronchitis. Again, the poor results seem indicative of sampling error. Less than 10 % of the variation across SMSAs is explained by regression 35. No pollution variable appears significant. Of the socioeconomic variables, the proportion of poor people is significant and there is an indication that the number of old people in an area is of some importance.

Unlike most of the previous results, when the occupation variables are added in the next regression (36), the pollution variable which was insignificant becomes quite significant. The poor variable also becomes more significant. Again, we see the unemployment rate related positively to the death rate and the white-collar category of labor related negatively to the mortality rate. In addition, the percentage of males in the labor force is associated negatively with bronchitic deaths, while employment in wholesale and retail trade as well as educational services is related positively to such deaths.

Summary and conclusion

We have examined the relationship between air pollution and mortality in 117 U.S. cities in 1960. In previous studies, we have reported the effect of different functional forms (Lave, 1971), of disaggregating the mortality rates by age, sex, and race and of adding additional explanatory variables representing the types of home heating equipment, home heating fuel, water heating fuel, air-conditioning, and climatology of.the areas. In this study we considered the effect of adding variables measuring the occupation mix in the city.

The conclusion of these various studies is that air pollution has a marked effect on the mortality rate. Using different functional forms or adding additional explanatory variables does not change the basic implication. Some of the basic results were cited in the text. For example, a 10 percent decrease in the measured level of both particulates and sulfates is estimated to decrease the total death rate by 0.9 %.

Elsewhere, we estimated that a 50 % reduction in air pollution would lower the economic cost of morbidity and mortality by 4.5 % (Lave & Seskin, 1970). Some perspective of the magnitude of this effect is seen by noting that a complete cure for all cancers would lower the economic cost of morbidity and mortality by 5.7 %. Another way of estimating the effect is to look at the possible increases in life expectancy that might be achieved by pollution abatement. A 50 % reduction in air pollution is estimated to increase the life expectancy of a new born infant by 3 to 5 years. In view of the fact that the advancements in medicine and the delivery of medical care in the last two decades have not succeeded in raising the life expectancy (in the U.S.) at all, this is a noteable contribution.

This study is one additional piece of evidence documenting the association between mortality and air pollution. The estimated effects of air pollution on the mortality rate is quite consistent and important across different data sets and statistical models. Mortality rates could be lowered and life expectancy raised substantially by abating air pollution; it is time to implement a program of substantial pollution abatement.

References

1. Air quality criteria for sulfur oxides. *U.S. Public Health Service. National Air Pollution Control Administration Publication No. AP-50*, 1970.

2. Analysis of suspended particulates, 1957–61, *U.S. Public Health Service Publication No. 978*, 1962.

3. Anderson, D.: The effects of air con-

tamination on health. *Canadian Medical Association Journal 97*, 528–36, 585–93, 802–806 (in three parts), 1967.

4. Battigelli, M.: Sulfur dioxide and acute effects of air pollution. *Journal of Occupational Medicine 10*, 500–11, 1968.
5. *County and City Data Book*, U.S. Department of Commerce Publication, 1962.
6. Greenwald, I.: Effects of inhalation of core concentration of sulfur dioxide on man and other mammals. *Archives of Industrial Hygiene and Occupational Medicine 10*, 455–75, 1954.
7. Lave, L.: Air pollution damage. In *Research on Environmental Quality* (ed. A. Kneese). Johns Hopkins Press, Baltimore, 1971.
8. Lave, L. & Seskin, E.: Air pollution and human health. *Science 169*, 723–732, 1970.
9. Lave, L. & Seskin, E.: Air Pollution, Climate, and Home Heating: The Effect on U.S. Mortality, working paper (1970a).
10. Lave, L. & Seskin, E.: A Statistical Analysis of the Association Between U.S. Mortality and Air Pollution, working paper (1970b).
11. Lave, L. & Seskin, E.: Does Air Pollution Shorten Lives?, working paper (1970c).
12. *Vital Statistics of the United States (1960)*, U.S. Department of Health, Education, and Welfare Publication, 1963.
13. World Health Organization: Health effects of air pollution. Seminar report, *Bulletin of the World Health Organization 23*, 264–74, 1969.

Part IV

OPERATIONAL MANAGEMENT MODELS

MODELS FOR INVESTIGATION OF INDUSTRIAL RESPONSE TO RESIDUALS MANAGEMENT ACTIONS

*Clifford S. Russell**

Research Associate, Resources for the Future, Inc., Washington, D.C., USA

Summary

This paper argues that, for purposes of residuals management decisions, it is necessary to go beyond models which concentrate on end-of-pipe treatment and a single receiving medium. A model is proposed which reflects the physical tradeoffs among residual forms and receiving media implied by the laws of conservation of mass and energy, and which includes opportunities for reducing residuals generation in production as well as conventional treatment alternatives. A brief example is presented of the model's application to a petroleum refinery. Finally, the place of such individual models in an overall regional residuals management model is discussed.

Theoretical models of residuals management generally assume knowledge both of functions relating reductions in discharges of residuals by industrial firms or municipalities, to costs incurred by the discharging unit, and of functions relating external damages to those discharges, either directly or through the medium of environmental transfer functions.[1] Such models have been used to provide elegant proofs of very general propositions, giving us valuable insight into residuals-environmental quality problems and economically efficient methods of dealing with them. It is no reflection on these theoretical results to note that we are very far from having, in practice, this kind of knowledge of costs and losses. Neither should it be considered a criticism to suggest that for practical work, in connection with planning and policy decisions, cost functions dealing simply with the reduction in discharges of one or another residual are of little assistance and may even be misleading.

A major reason for suggesting doubts about the familiar cost-of-discharge-

---

* The author is heavily indebted to Blair T. Bower whose work in the field of industrial water use inspired the original versions of these models. In addition, valuable criticism of early drafts has been received from Allen V. Kneese and Walter O. Spofford, Jr.
[1] "Residuals management" may strike the reader as an odd term, but we believe it is preferable to its more common, rough equivalent, "pollution control". Our major reason for this feeling is that the former term makes explicit that the residuals, or leftovers, from human production and consumption activities are at the heart of the problem. In addition, the term "management" implicitly recognizes that, in the absence of 100 per cent recycling (as in a space ship) and dependence on solar energy, some residuals will always be with us requiring disposal, so that the rational choice of forms, amounts, location and timing of discharges to the natural world is the problem to which we must address ourselves. (We shall provide below a more careful definition of "residual". For now, we note that this way of stating the problem has the additional advantage of moral neutrality as opposed to the connotation of evil implicit in "pollution".)

reduction approach is that it encourages us to ignore the lesson of the laws of conservation of energy and mass. Thus, we tend to focus on a single residual (or on a single receiving medium) and to assume implicitly that when we have reduced discharges of that residual (or to that medium) the material or energy no longer discharged has conveniently disappeared.[1] In fact, of course, most of the common methods of discharge reduction simply change the form of the residual or the medium to which the residual is discharged. Thus, for example, standard sewage treatment processes remove oxygen-demanding organics from waste-water by precipitation and biochemical decomposition, but produce a biologically active sludge which demands further treatment and ultimate disposal. Incineration of municipal trash and garbage reduces the volume of "solid waste" by oxidizing much of the "raw" material. This volume reduction, however, carries with it an *increase* in the total weight of residuals (due to the added oxygen) and the discharge of a considerable amount of waste (heat) energy to the atmosphere (from the exothermic combustion reaction).[2]

A second difficulty with applying the standard cost functions for discharge reductions is that they generally offer us no clue to what actions on the part of the discharging plants might lie behind them and, in consequence, lead us to concentrate on end-of-pipe treatment methods. In reality, a number of methods, which we shall discuss below, are available to most industrial plants for reducing their *generation* of residuals, and these are frequently cheaper than processes designed only to reduce specific discharges for given generation levels. Many of these methods involve, however, changes in the production process itself, and hence their adoption will depend not only on public residuals management policy, but also on changes in processing technology; relative input prices and available input qualities; and relative output prices and output quality requirements. Many of these are, in turn, influenced by policies of government seemingly unrelated to residuals management.[3]

---

[1] It will be useful at this point to be more precise about what we mean by a "residual". If we think of a production (or, indeed, a consumption) process as a black box into which flow inputs of mass and energy (as well as of factor services), out of that process will flow an equal amount of energy and an equal amount of mass (except for a tiny conversion of mass to energy in nuclear reactions). For a production process, one or more of the energy or mass outflows will constitute the product. The other flows will, in general, have zero prices in existing markets, or at least prices below their variable costs of production, transportation, etc. These are the process residuals and the firm's only concern is their minimum cost disposal. In consumption processes, of course, the output is the maintenance of human life, or the provision of "utility". Aside from the "investment" implied by a growing population, consumption residuals will, in the aggregate, equal inputs. And thus, for the economy as a whole, material and energy residuals will equal material and energy inputs to production and only recycling will actually reduce the magnitude of residuals requiring disposal. This view of the problem has been most carefully and persuasively stated by R. U. Ayres and A. V. Kneese in "Production, Consumption and Externalities", *American Economic Review*, June 1969, Vol. 59, No. 3, pp. 282–297.
[2] The combustion residuals include CO_2 and water vapor.
[3] An example of such a government policy with important implications for residuals discharges is the oil import quota system in the United States. This system influences the sulfur content of crudes charged in domestic refineries, thus affecting refinery wastes, and has encouraged the import of relatively high sulfur Venezuelan residual oil for use in heating and electric generation on the East Coast.

In view of the above considerations, it seems advisable that for practical investigation of industrial response to residuals management policies we broaden our approach to reflect both the interconnections between residual forms and discharge media implied by the conservation of mass and energy, and the opportunities available for the reduction of residuals generation in production. As a first step in this direction, we suggest a description of the firm's decision-making process concerning residuals discharges. We then modify this description for applied work by converting it to a form solvable by linear programming algorithms. A significant portion of the paper will deal with the techniques used to include residual generation, transformation and discharge as decision variables in such linear models. In addition, we shall point out some of the problems with our method, before briefly illustrating its use with a model of a petroleum refinery.

The Conceptual Model

The fundamental idea behind the conceptual model is that we can fruitfully look at industrial residuals generation and discharge patterns as resulting from a two-level decision-making process within the firm.[1] At one level are the decisions surrounding the basic production process itself and the determination of input mix and scale of activity such that the plant's contribution to profit is maximized. At this level we identify gross residuals generation which is assumed to be treated symmetrically with other inputs such as labor, capital and raw materials.[2] At a second level in the decision process, we conceive of an effort being made to minimize the costs associated with "supplying" the quantity of primary residuals generation decided on at the first level.[3] At this second stage, technology and prices associated specifically with by-product recovery, recirculation, preliminary treatment, transport within the plant, end-of-pipe treatment and discharge are explicitly considered. The resulting costs associated with the basic input might then be used in another round of calculation of the optimum of the first-stage model. The process

---

[1] Formally, the two-level process we describe could obviously be represented by a single non-linear programming problem. We choose the two-level model primarily for two reasons. First, we believe it is probably a reasonable description of how such decisions actually are made, in the sense that production considerations are the domain of one group which communicates its needs for "utility services" to a second group specializing in this area of plant planning. Second, the model makes explicit the link between residuals discharges and such considerations as product quality constraints and advances in process technology.

[2] Residuals constitute, as we have noted, material or energy flows out of the production process, but their generation may fruitfully be analyzed as an input analogous to other familiar inputs and substitutable for some of them. For example, in petroleum refining, capital, in the form of heat exchangers (which transfer heat from a hot output stream to a cool input stream) substitute for the generation of waste heat (which would have to be removed by cooling water or air) as well as for capital, in the form of furnace stills for heating the input, and for fuel burned in those furnaces.

[3] The initial solution to the first stage problem requires the use of some price set for residuals generation. We might think of this first approximation as being the cost of once-through use of all carrying streams with no treatment prior to discharge, or with just sufficient treatment to meet existing standards.

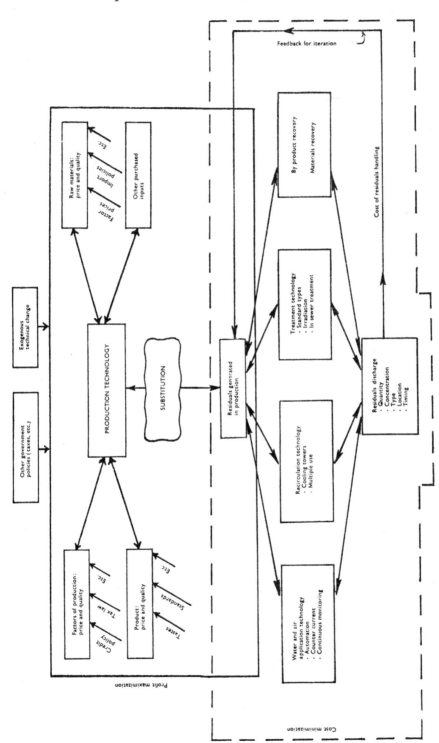

Fig. 1. A proposed model of industrial residuals generation and discharge

of iteration could continue until the costs of another round exceeded the likely increase in profits to be gained by further refinements. Fig. 1 provides a schematic view of the conceptual model.

This model is only a conceptual base for the working models which follow, but it seems even in this qualitative form to serve two useful purposes. First, it focuses attention separately on the influences outside the firm which indirectly and those which directly affect residuals generation patterns. And second, it emphasizes that within the production process itself other inputs can frequently be substituted for primary residuals generation. That is, it illustrates the fundamental sense in which it is misleading to assume that analyses based on fixed coefficients (such as pounds of *BOD* generated per unit of output) are conceptually valid.

The Working Models

For quantitative analysis of industrial response to residuals management actions, the conceptual model briefly outlined above is of limited usefulness. This is so primarily because of the great difficulties in identifying the form and estimating the parameters of the basic neo-classical production functions at the heart of the first phase optimization.[1] To avoid this problem, our approach in constructing working models is to attempt to identify only some relatively small set of discrete production alternatives, to cast these in the form of unit activity vectors (that is, vectors giving inputs and cost required for the production of one unit of the output of interest), and to condense the two phase decision process into a single phase represented by a linear program.[2] The objective of the firm may be taken to be profit maximization, cost minimization for given output, or any other convenient variant. The constraint set may include limits on input availability; quality requirements to be met by products; limits on discharges of one or more residuals; and, most importantly, continuity conditions (or mass and energy balance equations) requiring, for example, that the full amount of each residual generated be accounted for explicitly either by intake to a treatment or transport process

---

[1] These difficulties are of three sorts: (*i*) the heterogeneity of the mixes of sub-processes making up various observed production processes which seem to be homogeneous if we look only at raw materials and outputs; (*ii*) a lack of data on output and the associated quantities of various inputs for a range of relative input prices, even where reasonably homogeneous processes may be postulated; (iii) statistical inference problems in the use of whatever data may be available. (On the last point, see for example, I. Hoch, Simultaneous Equation Bias in the Context of the Cobb-Douglas Production Function, *Econometrica*, October 1958, 566–78.)

[2] Models of this sort, dealing only with waterborne residuals are described for thermal electric generation, beet sugar manufacture and petroleum refining in C. S. Russell, "Industrial Water Use", Section II of a Report to the National Water Commission by RFF, "Future Water Demands: The Impacts of Technological Change, Public Policies and Changing Market Conditions on the Water Use Patterns of Selected Sectors of the U.S. Economy: 1970–1990", June 1970.

or by discharge.[1] (As we describe below, secondary residuals are generated in these treatment and transport processes and are subject to the same continuity conditions.)[2]

The general form of the model is shown schematically in Fig. 2 and may be interpreted as a familiar linear programming problem:

max $c'x$

subject to $Ax \leqslant b$

and $x \geqslant 0$.

The activity levels (x) which are to be chosen by the solution process are shown across the top under six headings: production alternatives ($X_1 \ldots X_H$); by-product extraction ($B_1 \ldots B_J$), and raw material recovery ($W_1 \ldots W_K$) from residuals; treatment and transport of residuals ($T_1 \ldots T_L$, $V_1 \ldots V_M$); discharge of residuals ($D_1 \ldots D_G$); and sale of products ($Y_1 \ldots Y_N$). Corresponding to this division of the possible activities is the vertical division of the A-matrix and the objective function. Thus, for example, the objective function entries corresponding to the production activities are the costs of production; those corresponding to residuals discharges are zero unless some effluent charge is being levied.[3]

The horizontal division of the A-matrix indicates broadly the type and form of the constraints included. Thus, each unit of production using process X_1 requires a vector of inputs represented by ($-\mathbf{p}_{X_1}$). Total input requirements for production at level X_1 are $-\mathbf{p}_{X_1} \cdot X_1$, and the input availability constraints simply say that

$$\sum_h (-\mathbf{p}_{X_h}) \cdot X_h \geqslant -P,$$

or that no input may be used beyond its level of availability as given by the applicable entries in the right-hand side or b-vector.[4] The other constraints

---

[1] To say that once a quantity of residual is generated in the model, it must be treated, transported or discharged, is not to say that *all* residuals must be included in the model. For many purposes, for example, the CO_2 and H_2O residuals from combustion processes will not be of interest and may be ignored in the construction of a response model. Which residuals are of interest will depend to a large extent on the spatial and time dimensions of the study. Thus CO_2 may be a problem on the global, long-term scale, but is generally not so considered for local areas over the next decade or so.

[2] We use the term "secondary residuals" to refer to residuals generated in such auxiliary processes as treatment, recirculation and byproduct extraction, as opposed to those generated in the production process itself (the "primary residuals"). This distinction is made for convenience only. (See below, pages 148 and 149 for further discussion of secondary residuals.)

[3] In this paper we concentrate on the features of the model specifically designed for dealing with residuals generation and subsequent handling, but for real industrial applications, the section including alternative production processes may be extremely complex in itself. See, for example, the diagram and description below of the oil refinery to which we applied these techniques. This part of the problem is, however, covered in detail in other sources. See Alan S. Manne, *Scheduling of Petroleum Refinery Operations* (Cambridge: Harvard University Press, 1963), for one industry.

[4] The minus sign on inputs is used here primarily to emphasize the symmetry between traditional inputs and primary residuals generation. In practice, signs may be chosen for computational ease (for example, so as to avoid negative entries in the b-vector), so long as row consistency is maintained.

Columns ... / Rows	Production alternatives $X_1 \ldots X_H$	Byproduct extraction $B_1 \ldots B_J$	Raw material recovery $W_1 \ldots W_K$	Treatment and transport of residuals $T_1 \ldots T_L \ldots V_1 \ldots V_M$	Sale of products $Y_1 \ldots Y_N$	Discharges of residuals $D_1 \ldots D_G$	Right hand side
Production and sale	$a + ex_1 \ldots + ex_H$	$+ b_1 \ldots + b_J$			$-ey_1 \ldots -ey_N$		$\geqq 0$
Input availability	$-px_1 \ldots -px_H$		$+w_1 \ldots +w_K$				$\geqq -P$
Output quality	$+qx_1 \ldots +qx_H$						$\geqq Q$
Primary[b] residuals	$-rx_1 \ldots -rx_H$	$+eB_1 \ldots +eB_J$	$+ew_1 \ldots +ew_K$	$+eT_1 \ldots$ $+ev_1 \ldots$		$+eD_1 \ldots$	$= 0$
Secondary[b] residuals		$-rB_1 \ldots -rB_J$	$-rw_1 \ldots -rw_K$	$-rT_1 \ldots +eT_L \ldots +ev_M$ $-rv_1$ $-rT_L$ $-rv_M$ Etc.		$\ldots +eD_g \ldots = 0$ . . . $+eD_G = 0$	$= 0$
Possible discharge constraints						$+1 \ldots +1 \leqq F$	
Objective function	Costs of production	Costs of extraction	Costs of recovery	Costs of treatment and transport	Prices of output	·Possible effluent charges	

[a] The **e** are column vectors of zeroes and ones. A particular vector **e** has the number of row elements corresponding to the constraint set in which it appears. The occurrence of ones is determined by the function of the column in which the vector appears. Thus, in ex_1, a one appears in the row corresponding to the output of process X_1.

[b] For further discussion of the structure of these constraints, see the text.

Fig. 2. Schematic of models of industrial residuals generation, treatment and discharge

imposed include requirements that all that is sold be produced; that output quality conform to certain standards;[1] and that all residuals generated be accounted for by raw material recovery, byproduct extraction, treatment, transport or discharge. This last requirement deserves more detailed discussion.

The choice of a treatment or transport option for dealing with some residual generated in production implies the generation of one or more secondary residuals.[2] These are basically of three types: that portion of a residual subject to treatment which is not removed or altered by the treatment process (as the fly ash continuing up the stack after electrostatic precipitation); the new forms of residuals created by a treatment process (as the sludge from primary and secondary biological treatment of waterborne oxygen-demanding compounds, and CO_2 from refuse burning); the same residual at a different *place* (as when sewage is piped elsewhere before discharge). The method of inclusion of secondary residuals and of the requirement that they, in turn, be treated, transported or discharged is basically straightforward and might be called "row transfer". For example, consider an activity, say T_l, designed to remove some percentage, b, of residual i from the stream containing it, while in the process generating a quantity of residual, say $r_{Tl'}$, all at a cost c_{Tl} per unit of i taken in to the process (*not*, in this method, the cost per unit of i removed). The activity vector for this process would be:[3]

Row description	Treatment Activity (T_l) vector entries
residual r_i generated in production and to be accounted for	$+1$
residual r_i *not* removed	$-(1-b)$
new residual generated per unit intake	$-r_{Tl'}$
objective function	$-c_{Tl}$

---

[1] We have shown the quality constraints for production at levels $X_1 \ldots X_H$ in the simple form

$$\sum_h qx_h \cdot X_h \geqslant Q.$$

Because such constraints will generally be placed on the concentration of one or another substance in the product, and because final products often result from blending separate intermediate stocks, the actual construction of these rows will generally be more complicated. The method we discuss below for including effluent concentration constraints may be applied to such product quality requirements.

[2] As we show, byproduct extraction and raw material recovery will also give rise to secondary residuals. The techniques for including such options is generally the same as that described in the text except for the necessary entries reflecting the recirculated stream (e.g., cooling water) or the recovered material.

[3] We concentrate here on tracing residuals flows, but the treatment and transport processes will also require *inputs*, such as steam, water, chemicals and electricity, which will, in turn, influence directly the plant's generation of residuals. These inputs must also be included in the activity vectors in order that these influences be felt. Including only the costs of such inputs would not allow reflection of their direct effects on residuals loads.

	Residuals generation	Activity levels		
		Treatment	Transport	Discharge

$$X_f \ldots X_k \ldots \quad T_l \ldots T_m \ldots \quad T_n \ldots V_o \ldots \quad D_i \ldots D_{i'} \ldots D_j \ldots D_h \ldots D_g$$

row i: $-rx_{fi} - rx_{ki} .. +1 \ldots +1 \ldots \quad +1 \ldots$

row i': $\ldots -(1-b_l) .. -(1-b_m) \ldots \quad +1 \ldots$

row j: $-rT_{lj} \ldots -rT_{mj} .. +1 \ldots +1 \ldots \quad +1 \ldots$

row h: $-rT_n h \ldots \quad +1 \ldots$

row g: $-rV_o g \ldots \quad +1$

$$= 0$$

Obj. Fn. $\quad -cx_f - cx_k \quad -cT_l - cT_m \quad -cT_n - cV_o \quad * \quad * \quad * \quad * \quad *^a$

[a] The asterisks indicate that prices (effluent charges) may be applied to the discharges activities.

Fig. 3. Residuals handling in the linear model

Thus, if this treatment activity is operated at the level, T_l, sufficient to account for all the residual i assumed generated, the quantity $(1-b)T_l$ of i would be unaffected, requiring further treatment or discharge; and the quantity $rT_{lj} \cdot T_l$ of the new residual j would now also have to be accounted for. But let us follow this hypothetical residual, i, through the constraint matrix, using the notation of Fig. 2. In Fig. 3 we show the necessary matrix entries, the activity level designations, the right hand side and the objective function.

Treatment activities "l" and "m" remove, respectively, the fractions b_l and b_m of the amount of residual i to which they are applied. Thus, for every unit of i taken in to process "l", $(1-b_l)$ units remain for discharge; assuming, as we have here, no opportunity for further treatment. Also, for every unit of i taken in, the treatment processes produce, respectively, rT_{lj} and rT_{mj} units of a new residual, "j". This, in turn, must be treated in process "n", transported by process "o" or discharged. And finally, both processes "n" and "o" produce further residuals requiring discharge.[1] Writing out the individual row conditions for this example, we have:

$$-X_f \cdot rx_{fi} - X_k \cdot rx_{ki} + T_l + T_m + D_i = 0$$
$$-T_l(1-b_l) - T_m(1-b_m) + D_{i'} = 0$$

---

[1] Usually a transport activity simply changes a residual's location and not its form or amount. Thus the entry $-rv_oG$ might just as well be equal to -1. This may strike the reader as trivial in a model of a single plant, but the distinction between discharge locations is far from trivial in the broader context of regional residuals management decisions. (See Concluding Comments, below.)

$$-T_l \cdot r_{Tlj} - T_m \cdot r_{Tmj} + T_n + V_o + D_j = 0$$
$$-T_n \cdot r_{Tnh} + D_h = 0$$
$$-V_o \cdot r_{Vog} + D_g = 0$$

In the absence of effluent charges, the contribution to the objective function of this section of the problem is given by:

$$-X_f \cdot c_{X_f} - X_k \cdot c_{X_k} - T_l \cdot c_{T_l} - T_m \cdot {}_{T_m} - T_n \cdot c_{T_n} - V_o \cdot c_{V_o}.$$

One extension of this general method is worth specific mention, for it allows us to deal rather efficiently with the common situation in which several streams (as, for example, process water streams from different processing units) contain a number of residuals in different proportions and are subject to several possible treatment states, each removing a particular proportion of each residual. The basis of this method is to use, in place of any single residual quantity, the quantity of the carrying stream as the variable in the required row continuity conditions. Only at the point of discharge are the concentrations of residuals used to obtain quantities of residuals explicitly. Thus, assume we are interested in two water streams of volumes V_1 and V_2, generated in production processes X_1 and X_2 respectively. Assume that both contain three residuals of interest, in concentrations O_{11}, O_{12}, O_{13} and O_{21}, O_{22}, O_{23} respectively. Further assume that each stream may be discharged directly; subject to treatment process A and then discharged; or subject to process A followed by process B and then discharged. Process A is assumed to remove fractions a_1, a_2 and a_3 of the three residuals; proces B removes fractions b_1, b_2 and b_3.[1] We neglect, for simplicity, all process inputs and any secondary residuals generated in A and B. Then the required matrix entries, etc., may be written as in Fig. 4, and the total discharged quantities of residuals 1, 2 and 3 are simply obtained as the sum of discharges 1 through 6.[2]

The most obvious problem with this approach involves economies of scale. If any subset of the several streams may be chosen for treatment and others discharged, we can never know in advance the total volume for which the facility must be designed. Hence unit costs applied to A and B in the objective function must necessarily be arbitrary.[3]

[1] The assumption that removal "efficiencies" are independent of relative and absolute concentrations may be unacceptably inaccurate in certain cases (as BOD and phenols or BOD and turbidity), but in such cases, the linear framework itself probably breaks down.
[2] This may be done by introducing three new activities D_7, D_8, D_9 with $+1$ entries respectively in the rows for residual 1, residual 2, and residual 3, and by requiring that the sum across each of those rows be zero. D_7 will then be the total discharge of residual 1, and so forth.
[3] We may guess at the probable volume, choose the largest possible volume, or adopt some other strategy, but whatever we do will have some effect, in turn, on the volume actually determined in the solution. As long as there are economies of scale, these guesses will tend to be self-fulfilling, because large assumed volume will imply low unit costs which will in turn encourage the wide adoption of treatment (and vice versa). More subtle problems may arise because of non-convexities introduced by variations in the removal per unit cost between A and B for one or more residuals. These are discussed more fully in Russell, op. cit., and are closely related to similar problems discussed below in this paper.

Activity levels

Rows	A_1^a	A_2	B_1	B_2	D_1	D_2	D_3	D_4	D_5	D_6
Vol. 1 (from production)	+1				+1					
Vol. 1 (from process A)	−1		+1			+1				
Vol. 1 (from process B)			−1				+1			
Vol. 2 (from production)		+1						+1		
Vol. 2 (from process A)		−1		+1					+1	
Vol. 2 (from process B)				−1						+1
Residual 1					$-O_{11}$	$-(1-a_1)O_{11}$	$-(1-b_1)(1-a_1)O_{11}$	$-O_{21}$	$-(1-a_1)O_{21}$	$-(1-b_1)(1-a_1)O_{21}$
Residual 2					$-O_{12}$	$-(1-a_2)O_{12}$	$-(1-b_2)(1-a_2)O_{12}$	$-O_{22}$	$-(1-a_2)O_{22}$	$-(1-b_2)(1-a_2)O_{22}$
Residual 3					$-O_{13}$	$-(1-a_3)O_{13}$	$-(1-b_3)(1-a_3)O_{13}$	$-O_{23}$	$-(1-a_3)O_{23}$	$-(1-b_3)(1-a_3)O_{23}$
Obj. Fn.	$-c_A$	$-c_A$	$-c_B$	$-c_B$						

^{a} A_1 indicates the level of process A applied to stream 1.

Fig. 4. Method of handling multi-residual streams subject to a variety of treatment processes

Having discussed in some detail how the linear response model reflects residuals handling alternatives, we now go on to describe how it can be used to investigate industrial response to residuals management action on the part of public authorities. We may use effluent charges, discharge quantity constraints or even discharge concentration limits to determine the effects of these direct measures on discharges, costs, production volume, etc. At another level, we may vary process technology, requirements for output quality, available input quality, relative input prices, etc., and under each set of these indirect influences, examine the impact of direct residuals management actions. Thus, if we have good reason to expect changes in any of the indirect influences, we are in a position to investigate the residuals generation and discharge pattern of the firm under the new conditions as well as under present conditions.

We have already indicated, in Fig. 3, the place of effluent charges in the model. Since discharges are explicit activities and not simply "slacks," effluent charges fit perfectly as the unit activity costs. If effluent charges, specific to residuals and to locations, were applied to the discharges in Fig. 3, the new objective function would include:

$$-(D_i+D_{i'})\alpha_i - D_j\alpha_j - D_h\alpha_h - D_g\alpha_g.{}^1$$

where α_i is the fee per unit of discharge of residual i. If the charges were not specific to location as well as type, we would have:

$$-(D_i+D_{i'})\alpha_i - (D_j+D_g)\alpha_j - D_h\alpha_h$$

since, by hypothesis, residual "g" differed from residual "j" only in discharge location. Constraints on discharge quantities are easily included by attaching additional rows. If we wish to constrain a specific discharge D_i, to be less than $\bar{D}_i$, we simply introduce a new row in which the activity (column) D_i has a $+1$ entry. Then we put a constraint on the new row, requiring $D_i \cdot 1 \leqslant \bar{D}_i.{}^2$ If we wish to constrain discharge concentrations (as mg of BOD per litre of water), our task is somewhat more complicated. Referring back to Fig. 4, consider the possibility of constraining the concentration of residual 1 to be less than $\bar{R}_1$. The total discharge of residual 1 is

$$Q = O_{11}D_1 + (1-a_1)O_{11}D_2 + (1-b_1)(1-a_1)O_{11}D_3 + O_{21}D_4$$
$$+ (1-a_1)O_{21}D_5 + (1-b_1)(1-a_1)O_{21}D_6$$

---

[1] Throughout this paper we ignore the dimension of time, but in a practical application the time pattern of discharges may be very important because of variation in the assimilative capacity of the natural world. Then options to store residuals for later discharge and on–off options for treatment operation become important, and effluent charges may be differentiated by timing as well as type and location. These dynamic considerations add considerably to the complexity of the model and require different solution techniques.

[2] If several separate discharge activities all involve the same residual type, it is a simple matter to constrain their sum.

while the total volume of the discharge stream is $D_1 + ... + D_6$. Thus, the concentration is

$$\frac{Q}{D_1 + ... + D_6};$$

and we wish to require that:

$$\frac{Q}{D_1 + ... + D_6} \leqslant R_1.$$

In this form, the constraint is unusable in the linear program, but if we clear fractions and subtract the right hand from the left hand side, we obtain:

$$D_1(O_{11} - R_1) + D_2[(1 - a_1)O_{11} - R_1] + ... + D_6[(1 - b_1)(1 - a_1)O_{21} - R_1] \leqslant 0$$

Thus, the row entries become the differences between the actual and the desired concentrations, and the constraint simply says that the volume-weighted average of these differences must be less than or equal to zero.[1]

A variety of indirect influences on residuals generation and discharge may also be studied by manipulation of values of the objective function, the right-hand side, or the matrix of coefficients itself. Thus, in the objective function any of the price or cost figures may, in principle, be altered and the effect observed, though in practice we may be interested only in the price of a key input (such as coal to a thermal electric generating plant) or of a particularly important product (such as motor gasoline from an oil refinery). On the right-hand side, we may change input availabilities and output quality requirements. And, finally, we may reflect advances in production or waste handling technology by changing coefficients within the A-matrix itself. Such changes may take the form of introducing entire new columns to represent possible new processes. Another alternative is to change one or two coefficients within existing columns to reflect progress in some aspect of a largely unchanged process.

Some Difficulties

In the above description, we have already mentioned several potential sources of difficulty in our techniques for constructing linear models of industrial production, residuals generation, treatment and discharge. One of these was the matter of scale economies; which arises any time unit capital or operating costs vary inversely with the scale of the facility. On the practical level, there is no single correct unit price to attach to the activity vector for such a facility,[2]

---

[1] This method has been adapted from a Harvard Business School discussion problem: "The Tascosa Refinery", ICH 9C47R1, TOA, 4R2, Harvard University, 1965. There it was used to set lead concentration limits on gasoline blended from several stocks.
[2] That is, there is no single correct price until after the problem has been solved and the scale of the facility is known.

and the usual trick of approximating the nonlinear curve with piecewise linear segments will not work because the segments would not fill up in the correct order in the solution process. But, on a more fundamental level, economies of scale make any problem involving a choice of capacity a difficult one because the response surface, corresponding to the falling marginal and average costs, will have multiple optima.[1]

Similar effects are created whenever the marginal cost of obtaining some desired result fall as the amount obtained rises. Thus, in standard sewage treatment for removal of oxygen-demanding organics, the cost of BOD removal, when graphed against percentage removal achieved, follows an S-shaped curve, with falling marginal costs of additional removal over a significant range.[2] Thus, the determination of the appropriate treatment level is subject to the same difficulties as is that of the proper size of a facility exhibiting economies of scale. It is possible to set up the constraint matrix entries for a standard treatment plant in such a way that the segments must be chosen in the proper order,[3] but the problem will still involve non-convexity.[4]

A related problem arises if we wish to model the situation in which we are to decide whether or not to install some residuals treatment equipment and, if it is installed, whether or not to operate it. One may again think of declining marginal cost: a first very steep cost segment reflects the installation, but not operation, of the equipment; then a flatter segment reflects operation costs (once installed) up to capacity (see the figure on page 148).

This problem may be approached through integer or dynamic programming solution methods although there is a practical limit in both methods to the number of such alternatives that may be considered.

---

[1] In a general programming framework, it is at least conceptually possible to deal with the multi-peaked surface by making random starts within the feasible space. See Peter Rogers, "Random Methods for Non-convex Programming", Doctoral Dissertation, Division of Engineering and Applied Physics, Harvard University, 1966.
[2] See, for example, Richard Frankel, "Economic Evaluation of Water Quality. An Engineering-Economic Model for Water Quality Management", University of California, College of Engineering and School of Public Health, SERL Report 65-3, January 1965.
[3] See, for example, C. S. Russell and Walter O. Spofford, Jr., "A Quantitative Framework for Residuals Management Decisions", presented at the RFF Conference "Environmental Quality and the Social Sciences: Theoretical and Methodological Studies", Washington, D.C., June 16–18, 1970; and for a different approach, D. P. Loucks, "Stochastic Methods for Analyzing River Basin Systems", OWRR Project C-1034, Technical Completion Report, Cornell University, Department of Water Resources Engineering and the Water Resources and Marine Sciences Center, August, 1969.
[4] As we have mentioned above, removal efficiencies, where more than one residual is involved in a *single* process, may be interdependent. (For example, BOD and phenols removal in a standard biological treatment plant.) This interdependence essentially introduces cross-product terms into the problem and makes the linear model of doubtful value. If, however, interdependencies exist only *between* processes, the problem is potentially amenable to solution. Thus, the removal efficiency of electrostatic precipitation of particulates in stack gases is directly related to the SO_2 content of those gases. Hence, burning low sulfur coal in a thermal electric plant will reduce sulfur oxide emissions, but may increase particulate emissions for the same equipment and operating policy. This sort of interdependence can be handled within the linear framework through careful definition and linkage of relevant alternatives.

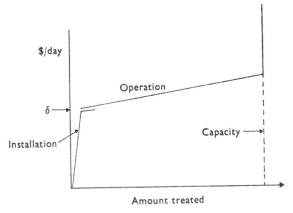

where δ→0

A Brief Example

As part of another study, the linear residuals response model was applied to the waterborne residuals of three different industries.[1] The most instructive of these applications was that to a moderately complex petroleum refinery, for this provided plenty of scope for studying the sensitivity of residuals generation and discharge to several of the direct and indirect influences discussed above.[2] By way of illustration, let us review this application briefly, describing the refinery and the residuals chosen for study, indicating specifically what sensitivity tests were made and summarizing a few of the results.

The petroleum refinery we modelled was intended as a relatively simple example of a unit designed primarily for the production of gasoline from original crude stocks. To this end, we included an array of processes for increasing the yield and quality of gasoline stocks per barrel of crude charged. We did not include any processes involved in lube oil manufacture, nor did we deal with petrochemicals except to assume that certain refining products might be sold as petrochemical feed stocks. Processing of non-gasoline products was kept to a minimum consistent with maintaining some reasonable range of choices at each stage. The refinery is depicted schematically in Figs. 5 and 6. In Fig. 5, we show the various possible petroleum processing units and the links between them, indicating the points at which the model had to choose between alternative courses of action.[3] Similarly, in Fig. 6, the flows of cooling

---

[1] See C. S. Russell, op. cit., "Industrial Water Use".

[2] Unfortunately, because this model was set up to aid in a study of industrial water use, we did not include the links between forms of residual and receiving media we have been discussing. Currently we are working to expand the model to include airborne and solid residuals.

[3] For certain important process units, the model also had a choice between different mixes of capital (in the form of heat exchangers) and fresh heat input/heat residual.

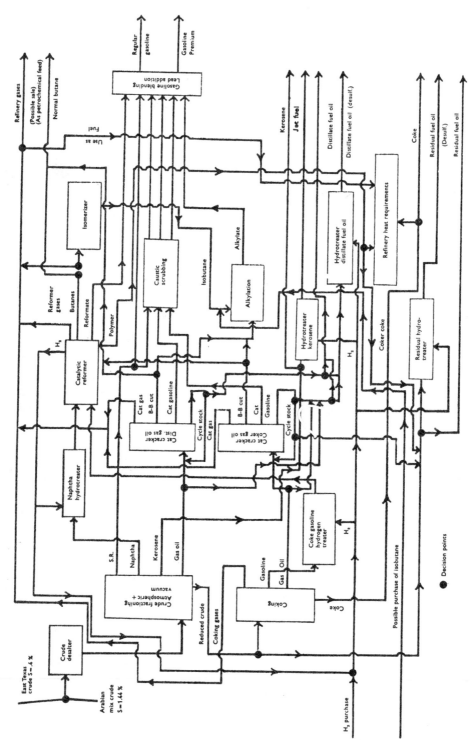

Fig. 5. Sample refinery for investigation of water use/waste generation schematic of product flows

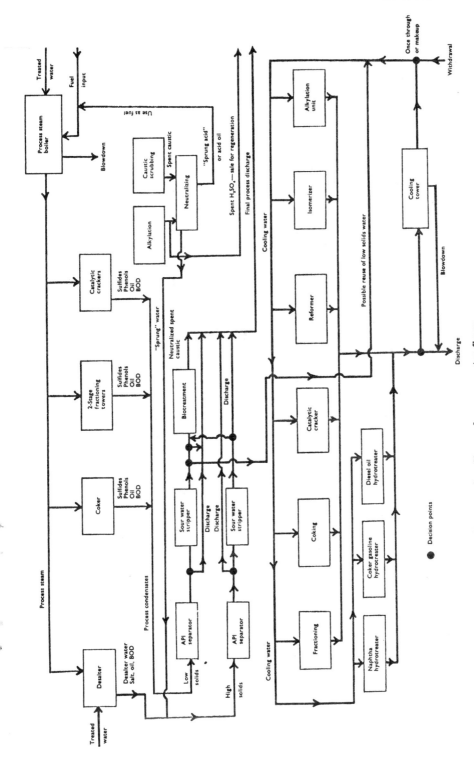

Fig. 6. Sample refinery — schematic flow diagram of cooling and process water flows

Table 1. *Influences investigated: petroleum refinery residuals generation and discharge*

I. Direct

A. Cost of cooling water withdrawals
B. Effluent charges on heat, sulfides, oil and BOD

II. Indirect

A. Product Quality Requirements: lead (TEL) content and octane of gasolines
B. Product mix: increase in jet fuel production as a fraction of crude oil charged
C. Input quality: sulfur content of crude
D. Relative input costs: for fresh heat, isobutane, hydrogen
E. Technological Change: improved cracking catalysts; adoption of hydrogen-intensive refining methods.[a]

[a] Hydrogen-intensive refining processes allow the refiner greater flexibility than conventional methods in choosing the ultimate yields to be obtained from any given crude charged. In addition, they can be used to remove sulfur and other impurities from any fraction, including the heavy fuel oils which now face strict sulfur content limits in several large cities. Finally, the amount of sulfur removed from the oil and its concentration in the residuals from the hydrogen processes often makes it economical to recover sulfur rather than discharge the sulfur compounds to air or water.

and process water and the possible related treatment and recirculation activities are diagrammed.

The analysis was based on the assumption that the refinery had not yet been constructed, but that certain decisions had been made which would constrain its size and possible configurations. This approach represented a compromise which allowed us scope in the inclusion of alternative processes, but which did not require us to consider the question of optimal size in relation to market demands and scale effects on costs. In particular, we assumed that there was a constraint on the daily availability of crude oil at the refinery, limiting the size of the installation as a whole, although allowing some range of sizes for most of the individual processing units. We chose the overall limit on deliveries as 50 000 barrels per day.

We allowed the refinery to make use of either of two crudes—one a relatively low sulfur oil which we gave the characteristics of an East Texas crude; the other, higher in sulfur, was modelled after an Arabian mix oil from the Persian Gulf.[1] The higher sulfur crude was assumed to be significantly cheaper but

---

[1] The crude characteristics, especially fractions and sulfur percentages, were drawn with some adjustments from M. M. Stephens and Oscar F. Spencer, *Petroleum Refining Processes* (College Park, Penn.: Pennsylvania State University, 1956). The other basic sources used were: W. L. Nelson, *Petroleum Refinery Engineering* (New York: McGraw-Hill, 1949); W. L. Nelson, *Guide to Refinery Operating Costs* (Tulsa, Okla.: Petroleum Publishing Company, 1966); M. R. Beychok, *Aqueous Wastes from Petroleum and Petrochemical Plants* (New York: John Wiley, 1967); A. S. Manne, op. cit.; American Petroleum Institute, *Manual on Disposal of Refinery Wastes:* Volume on Liquid Wastes (New York, 1969); M. C. Forbes and P. A- Witt, "Philosophy Methods and Costs of Refinery Waste Disposal", National Petroleum Refiners Association, Tech. 65–19, June 1965; Bonner and Moore Associates, Inc., *U.S. Motor Gasoline Economics*, Vol. I, A study done for the American Petroleum Institute, New York City, 1 June 1967.

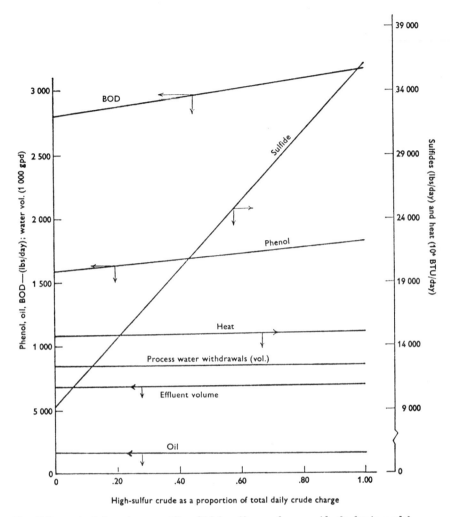

Fig. 7. Impact of changing quantity of high sulfur crude on residuals: basic model

subject to limited availability. This allowed us to explore the impact of crude oil sulfur content on refinery residuals.

The residuals traced through the model were heat, oil, BOD, phenols, and sulfides, and Table 1 summarizes the influences on the generation and discharge of these residuals which we investigated.

To give the flavor of the results obtained from such sensitivity tests, we present two examples. First, in Fig. 7 we show the variation in residuals generation (per day) with variation in the relative amounts of the high and low sulfur crudes charged. Naturally the sulfide generation changes most dramatically, but other residuals are also influenced by the change in crude oil type. Thus, BOD and phenols generated both increase as the charge of high-

Table 2. *Variation in generation of water-born residuals with changes in product quality and product mix: hydrogen-intensive refinery*[a]

All quantities are per barrel of crude oil charged

Residuals generated	Leaded gasoline allowed[b] Present octane requirements[c] No requirement on quantity of jet fuel produced[d]	No lead in gasoline Low octane requirements[c] No requirement on quantity for jet fuel produced[d]	No lead in gasoline Low octane requirements Jet fuel production = 17 % of crude oil volume[d]
Heat (10^6 BTU)	.364	.394	.379
Sulfide (lb.)	.264	.232	.094
BOD (lb.)	.058	.037	.018
Phenol (lb.)	.032	.020	.008
Oil (lb.)	.008	.008	.005

[a] A "Hydrogen-intensive Refinery" is one employing a number of refining processes in which hydrogen is introduced to the reaction vessel along with the liquid charge stock. Such processes may be used to remove sulfur and other impurities and to increase yields of motor gasoline or kerosene among other possibilities.

[b] "Lead" is tetraethyl lead (TEL) used to increase octane ratings of gasoline stocks.

[c] "Present Octane Requirements" taken to be: Premium (RON) = 100; Regular (RON) = 94. "Low Octane Requirements" taken to be: Premium (RON) = 92; Regular (RON) = 90, "RON" is research octane number.

[d] When no requirement is placed on the quantity of jet fuel produced, the assumed relative price set results in very little straight-run kerosene being upgraded to jet fuel. Straight-run kerosene (that resulting solely from fractional distillation of crude oil) is about 10 percent of crude volume. To obtain a 17 % yield of kerosene for jet fuel use, a hydrogen "cracking" process must be used. The yield of motor gasoline is correspondingly lower.

sulfur crude increases, because of the different characteristics of the Arabian crude.

A second example, this time of the sensitivity of residuals generation to changes in product quality specifications and product mix, is shown in Table 2. There, we note first the impact of prohibiting the introduction of any tetraethyl lead (TEL) to motor gasoline, while at the same time reducing the octane requirements for premium and regular grades.[1] Thus, the first and second columns indicate that these two product specification changes result in a larger waste heat load (because the octane upgrading processes necessary in the absence of TEL are heat intensive), and lower sulfide residual. The reduction in sulfide results from our assumption in the model that the concentration of sulfide in the processwater effluent from the hydrogen-intensive processes was great enough to warrant sulfur recovery. There is no significant change in the oil residual, and the small change in BOD and phenol loads is

[1] The present debate in the United States about the proper way to attack auto pollution focuses on the role of TEL as a problem in itself and as a fouler of catalytic devices for reducing emissions of CO and unburned hydrocarbons. It seems likely that the outcome will be lower-lead, lower-octane gasolines in the future.

to some extent an artifact of our assumption that water streams subject to the sulfur recovery process were stripped of these two residuals also. Our information also indicated, however, that generation of BOD and phenolic residuals was less in the hydrogen-intensive processes.

Similar effects to those noted above result when the product mix is changed to include considerably more kerosene per barrel of crude and to require the upgrading of all of this kerosene to jet-fuel specifications by hydrogen treating. Again, increased use of hydrogen-intensive processes is the basic response of the refinery. The slightly lower heat residual results largely from a decrease in the quantity of motor gasoline produced and the consequent lower level of operation of certain heat-intensive gasoline octane upgrading processes.

Concluding Comments

We believe that models of industrial response to residual-management action of the type described in this paper represent a useful tool for those entrusted with developing the technical basis for regional residuals management plans, for it permits the testing of different management actions, under a variety of assumptions about important indirect influences.

On a more systematic level, we may imagine a set of response models for the major industrial plants in a region forming part of an integrated residuals management model. Such a model would take account not only of the physical quantities of residuals discharged, but also of the action of the regional environment (its atmosphere and water courses) in diluting, transforming, transporting and accumulating those quantities and yielding the ambient residuals concentrations actually experienced throughout the region. Further, the model might include explicit allowance for the damages suffered by those exposed to these ambient concentrations.[1]

An example of such a model is shown schematically in Fig. 8.[2] Clearly, the place of the response models we have been discussing is in the left hand portion of the section labelled "Interindustry LP Model."[3] The diagram shows how the

---

[1] These receptors may be human beings or living and non-living objects of human interest; for example, damages result from the serious decline of predatory bird populations in the U.S. (in part, at least, due to persistent pesticide accumulation through the food chain) because some people have the survival of these species (or of all existing species) as an argument in their utility functions.

[2] The diagram is taken from C. S. Russell and W. O. Spofford, Jr., "A Quantitative Framework ...", op. cit. That paper discusses this overall framework in some detail and also presents illustrative results from a didactic version having a beet sugar refinery and a thermal electric generating plant as industrial residuals sources.

[3] This label is not meant to indicate that complete input–output models are included. For applied work, probably only the major dischargers need be modelled. *But*, any input–output links between these plants should be physically reflected in the model for the same reason that intraplant flows affecting residuals generation must be included explicitly. Market links (sales at market prices) between the plants will not reflect the external costs associated with the residuals discharges occasioned by the production for the sales. (See the footnote 3 page 148 above.)

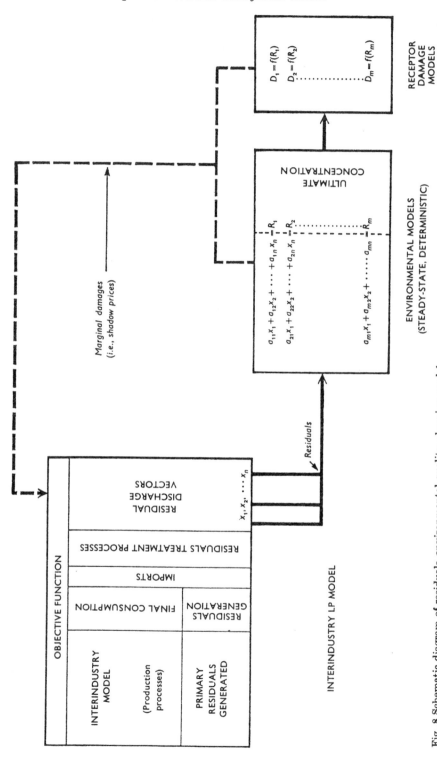

Fig. 8 Schematic diagram of residuals–environmental quality planning model

discharges resulting from solution of this submodel (with domestic sector attached) are traced through environmental models to yield ambient residuals concentrations.[1] These concentrations are, in turn, the arguments in receptor damage functions covering the range of direct and indirect effects on the people of the region.[2]

To illustrate the linkage between the three models and the method of solution, consider solving the linear response model initially with no restrictions or prices on residuals discharges. Using the resulting set of discharges, the environmental models and the damage functions, we may calculate the marginal damages attributable to each discharge; that is, the change in total damages which would result if that discharge were changed by a small amount. These marginal damages may then be applied as interim effluent charges on the discharge activities in the linear response model and that model solved again for a new set of production, consumption, treatment and discharge activities. If suitable provision is made for limiting the "distance" travelled from one iteration to another, this process will converge to a local optimum residuals management policy, including the effluent charge set necessary to achieve the specific features of that policy.[3] These results may be used to inform the relevant decision making unit whether it be an executive agency or a legislative body.[4]

Such an application of the response-model framework we have been discussing presupposes, of course, the availability of a considerable amount of detailed information about the specific plants in question. To the extent that such information is not available to the regional decision makers, all decisions will be made in conditions of uncertainty. It may, however, be possible to obtain approximately correct response information from general knowledge of the types of plants involved and models based on public information on that type of plant. Thus, the petroleum refinery model discussed above is a (simple) version of a standard type of refinery and was constructed using only publicly available data. Such an approach may be satisfactory in working on regional management models since our ignorance in other areas, such as damages attributable to health and esthetic effects is so great that refinement of the industrial response sub-model may seem a relatively low priority task.

---

[1] The actual form and method of derivation of these models is discussed elsewhere. See, for example, Russell and Spofford, op. cit. and Kneese's paper in this volume.

[2] The model may include, instead of or in addition to damage functions, constraints on the ambient residuals concentrations.

[3] See above for comments on non-convexities and the problem of multiple local optima.

[4] We may, however, wish to provide different kinds of information to these two types of bodies. See Russell and Spofford, op. cit. and C. S. Russell, "Regional Environmental Quality Management: A Quantitative Approach", a paper given at California Institute of Technology, Conference on Technological Change and the Human Environment, 21 October 1970.